14⁰⁰

Math
SIMULATIONS

Intermediate

Written by Karen P. Hall

Teacher Created Materials, Inc.
P.O. Box 1040
Huntington Beach, CA 92647
©1997 Teacher Created Materials, Inc.
Made in U.S.A.

ISBN 1-57734-192-3

Contributing Editor:
Janet Cain

Illustrator:
Howard Chaney

Cover Artists:
Sue Fullam
Chris Macabitas

Teacher Created Materials

Table of Contents

Table of Contents *(cont.)*

Introduction

Mathematics is a very important life skill. However, many students do poorly in this subject area because they do not understand why math is relevant to their everyday lives. As a result, they struggle with the concepts and skills. In order for students to think math is important and useful, they must understand its practical applications. *Math Simulations* helps students see how math is used in the real world. This progressive unit gives them something more to look forward to than the standard approach of drill and practice that is so commonly associated with math instruction. As students use the simulations, their interest and enjoyment of math will increase, causing their level of success to improve.

The 96 pages in *Math Simulations* are filled with a variety of fun-filled activities that focus on home economics. Students are given the opportunity to practice computational and problem-solving skills as they learn how incomes and expenses affect the finances and, ultimately, the lifestyles of different sizes of families.

The simulations in this book are organized as follows:

- Family Profile
- Professions
- Checking Accounts
- Estimated Family Budgets
    —Food
    —Housing
    —Transportation
    —Other Expenses
    —Unexpected Events
    —Total Expenses

Math Simulations is ideal for intermediate students. The easy-to-follow lesson design gives you maximum flexibility to adjust the curriculum to meet the needs of your students. Each lesson includes one or more objectives, suggested vocabulary to help present important concepts, a list of materials, and a detailed explanation of the procedure needed to complete the simulation. In addition, follow-up activities are frequently provided to extend the lessons. The last section of this book provides pages that allow students to practice and review a variety of mathematical skills. These activities can be utilized throughout the unit and are referred to in each lesson where appropriate.

Success with Simulations

The activities in *Math Simulations* have been selected in order to get students involved with math by actually simulating the conditions of a real-life application within the limited confines of the school environment.

Establish rules and procedures for organization that will remain consistent throughout each simulation. Suggestions for how to best utilize and store the units in this book are shown below.

Simulation Format

Simulations are organized according to the lesson plan format shown below. The necessary forms and/or playing cards are provided to make preparation for these activities as easy as possible for busy teachers.

> - Title of the Simulation
> - Objectives
> - Suggested Vocabulary
> - Materials
> - Procedure
> - Follow-Up Activities

Storing Simulations

As you use each activity, you will want to save the components of the simulation by using a readily available and well organized system that will serve your students in the future as well as the present. Labeled file folders or large manila envelopes can be easily sorted and organized by simulation units and kept in a file box. Pages that will be duplicated or made into overhead transparencies can be easily stored in the file folders or envelopes. Game cards and sample forms can be reproduced on heavy paper or cardstock and laminated. Then they should be placed in envelopes or resealable plastic bags before storing them in their respective folders.

Once the simulations have been organized into a file box, you will be prepared for each unit on a moment's notice. Students will enjoy having the opportunity to apply their math skills to real-world situations.

Cooperative Learning Groups

Cooperative learning is an important instructional strategy because it can be used as an integral part of many educational processes. It is made to order for activities that require higher-level thinking skills. The cooperative learning process acts as a powerful motivational tool.

Many of the activities in *Math Simulations* involve the cooperative learning process in a group effort to find solutions or come to conclusions. With this in mind, consider the following four basic components of cooperative learning as you initiate cooperative group activities.

Four Basic Components of Cooperative Learning

---

1. **In cooperative learning all group members need to work together to accomplish the task.** No one is finished until the whole group is finished and/or has come to consensus. The task or activity needs to be designed so that members are not each completing their own parts but are working together to complete one project.

2. **Cooperative learning groups should be heterogeneous.** It is helpful to start by organizing groups so that there is a balance of abilities within and among groups.

3. **Cooperative learning activities need to be designed so that each student's contribution to the group and each individual group member's performance can be assessed.** This can be accomplished by assigning each member a role that is essential to the completion of the task or activity. When input must be gathered from all members of the group, no one can go along for a free ride.

4. **Cooperative learning groups need to know the social as well as the academic objectives of a lesson.** Students need to know what they are expected to learn and how they are supposed to be working together to accomplish these goals. Students need to evaluate how well they worked to improve their social skills and to accomplish their academic objectives. Social skills are not something that students automatically know; these need to be taught.

Sample Lesson Plan

Each of the lessons suggested below and on page 8 can take from one to several days to complete.

Lesson 1	Lesson 2	Lesson 3
Introduce concept of simulations. Begin to fill out family profile (page 10). Pick family size (pages 11 and 12). Pick savings account (page 14). Draw picture of family (page 13).	Share family pictures (page 13). Introduce professions (pages 15 and 16). Pick professions (pages 17 and 18). Discuss classified ads. Find job in classified ads.	Discuss job applications. Introduce job application, using sample (page 19). Fill out job application (page 20).
Lesson 4	**Lesson 5**	**Lesson 6**
Discuss resumés (page 21). Introduce resumé format, using sample (page 22). Write rough draft of resumé (page 23).	Edit and proofread rough draft of resumé (page 23). Write final draft of resumé (page 23).	Discuss cover letters (page 24). Introduce cover letter format, using sample (page 25). Write rough draft of cover letter.
Lesson 7	**Lesson 8**	**Lesson 9**
Edit and proofread rough draft of cover letter. Write final draft of cover letter.	Hold job interviews.	Draw pictures of professions and write descriptions of duties.
Lesson 10	**Lesson 11**	**Lesson 12**
Discuss checking accounts (page 26). Introduce checks and deposit slips, using samples (page 27). Demonstrate how to keep a check register (page 29).	Practice writing checks and filling out deposit slips (page 28). Practice keeping a check register (page 29).	Predict family budgets (page 30).

Sample Lesson Plan *(cont.)*

Lesson 13	Lesson 14	Lesson 15
Discuss foods (pages 31–33). Introduce food pyramid (page 34). Review food prices (pages 35–38).	Discuss food menus. Introduce Sample Food Menu—Option 1(page 39) or Sample Food Menu—Option 2 (page 44).	Complete My Food Menu Option 1 (page 40) or Option 2 (page 45).
Lesson 16	**Lesson 17**	**Lesson 18**
Complete Food Menu summary—Option 1 (pages 41–43) or Food Menu Summary—Option 2 (page 46).	Discuss housing (pages 47–49). Examine housing price list (page 50). Fill out information about housing (page 51).	Discuss floor plans. Create floor plans (page 52).
Lesson 19	**Lesson 20**	**Lesson 21**
Discuss transportation (pages 53–55). Examine transportation price list (page 56). Complete transportation summaries (page 57).	Draw pictures of cars (page 58). Share pictures of cars.	Introduce/Explain other expenses (pages 59 and 60). Determine other expenses and compare product prices (page 61). Examine other expenses price list (page 62 and 63).
Lesson 22	**Lesson 23**	**Lesson 24**
Complete other expenses summaries (pages 64–66).	Discuss unexpected events (page 67). Pick unexpected events (pages 68–73).	Examine total expenses (pages 74–76). Determine if bills can be paid with designated income (pages 77–80). Discuss what was learned by using the simulations.

Family Profile

Objectives

Students will determine the sizes of their families and the amount in their savings accounts by randomly drawing activity cards. They will identify the types of expenses that all families have in common. They will begin to record financial information.

Suggested Vocabulary

adults, children, spouse, marriage, infant, cost of living, income, expenses, bills, recreation

Materials

- Family Profile Record Sheet (page 10), one copy per student
- Family Size (pages 11 and 12), enough for one card per student on cardstock or heavy paper
- My Family Picture (page 13), one copy per student
- Savings Account (page 14), enough for one card per student on cardstock or heavy paper

Optional:
- empty container

Procedure

1. Distribute the Family Profile Record Sheet to students. Explain that the purpose of this form is to keep a record of their incomes and to predict and estimate what their monthly and yearly expenses will be. Point out that this information is essential when determining a family's budget. (**Note:** The Family Profile Record Sheet will be used throughout this book. Students will continue to use their record sheets as they complete the remaining simulations.)

2. Discuss how family size affects income and expenses.

3. Cut apart the Family Size cards. Shuffle them and place them facedown in a stack or put them into an empty container. Invite each student to pick a card. Ask students to read aloud the information on their cards. Then tell them to record the information on their Family Profile Record Sheets.

4. Distribute My Family Picture to students. Have students draw pictures of their families according to the activity cards they have picked. You may wish to have them include the names, ages, and written descriptions for their family members. Encourage students to share their family pictures and information with the class.

5. Ask students to explain why savings accounts are important.

6. Cut apart the Savings Account cards. Shuffle them and place them facedown in a stack or put them into an empty container. Invite each student to pick a card. Ask students to read aloud the amounts shown on their cards. Then tell them to record the information on their Family Profile Record Sheets.

Family Profile Record Sheet

Name: _____

Number of Members in Family: _____

Members of Family: _____

Amount in Savings Account: _____

Income

Professions	Monthly Income	Yearly Income
Self:		
Spouse (if applicable):		

Predicted Budget   Estimated Budget

Expenses	Predicted Budget Monthly Amount	Predicted Budget Yearly Amount	Estimated Budget Monthly Amount	Estimated Budget Yearly Amount
Food				
Housing				
Transportation				
Other Expenses				
Unexpected Events				
Total				

Family Size

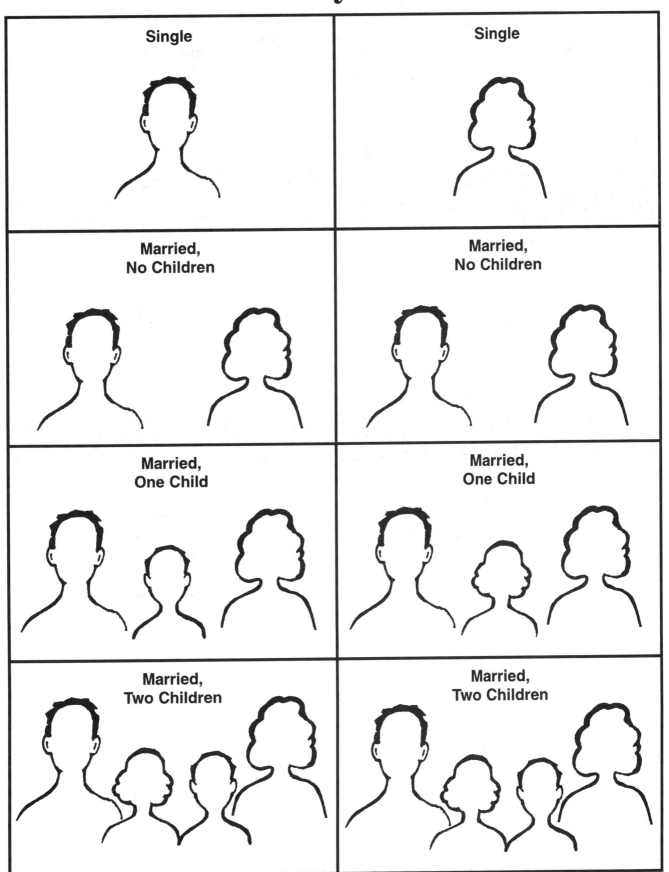

Family Size *(cont.)*

12

My Family Picture

Savings Account

Savings $100.00	Savings $500.00	Savings $5000.00
Savings $100.00	Savings $500.00	Savings $10,000.00
Savings $100.00	Savings $1000.00	Savings $10,000.00
Savings $250.00	Savings $1000.00	Savings $10,000.00
Savings $250.00	Savings $1000.00	Savings $25,000.00
Savings $250.00	Savings $5000.00	Savings $25,000.00
Savings $500.00	Savings $5000.00	Savings $25,000.00

14

Professions

Objectives

Students will pick professions and establish their incomes. They will learn how to fill out job applications, write resumés, and compose cover letters. They will use basic operations with whole numbers and add and subtract with decimals.

Suggested Vocabulary

want ads/classified ads, resumé, personal information, permanent address, education, experience, previous employment, work history, references, employer/boss, employee/worker, unemployed, self-employed, personnel department, application, applicant, citizen, male, female, interview, training, paycheck, social security number

Materials

- Professions and Salaries (pages 17 and 18), enough for two cards per student on cardstock or heavy paper
- Sample Job Application (page 19), transparency and one copy per student
- Job Application Form (page 20), one copy per student
- Writing a Resumé (page 21)
- Sample Resumé (page 22), transparency and one copy per student
- My Resumé (page 23), one copy per student
- Writing a Cover Letter (page 24)
- Sample Cover Letter (page 25), transparency and one copy per student
- Family Profile Record Sheet (page 10) from Simulation #1

Optional:

- empty container
- Basic Operations with Whole Numbers (pages 81 and 82)
- Understanding Decimals (page 83)
- Adding and Subtracting Decimals (page 84)
- Answer Key (page 93)

Procedure

1. Tell students that they are entering the job market. Cut apart the cards for Professions and Salaries. Shuffle them and place them facedown in a stack or put them into an empty container. Invite each student to pick two cards. Students should pick one of the two professions for themselves and use the other for their spouses if they are married. Students who are single (with or without children) should place the second card back in the stack or container. Have students record their professions and yearly incomes on the Family Profile Record Sheet (page 10).

2. Students can use the amounts shown on the cards or determine their incomes based on local classified advertisements for real jobs in the same career fields.

Professions *(cont.)*

Procedure *(cont.)*

3. Have students divide their yearly incomes by 12 to determine their earnings per month. Some students may find it helpful to use calculators.

4. Explain to students that the amount of money they earn often depends on the pay scale associated with a specific type of job, the level of education and training needed for the job, and how much experience they have had doing that type of work.

5. Ask students to brainstorm a list of characteristics that employers would be looking for in prospective employees. Discuss why each characteristic is important.

6. Introduce the employment section of the classified advertisements in newspapers. Invite students to read through the job listings. Each student should select one or more jobs for which she/he can apply. Students may need help with unfamiliar words and abbreviations used in the classifieds. For positions without stated salaries, tell students to use the average salaries shown on the cards they have picked.

7. Distribute the Sample Job Applications to students. Use the transparency to discuss the parts of this form. Give students the blank forms, and have them fill these out. Students should make up their past work experiences. Ask them to look at other classified job listings in the same career field, but for which they are not applying, to create this information.

8. Distribute the Sample Resumés to students. Use the transparency of the sample and Writing a Resumé to discuss the different parts. Distribute the blank resumés, and have students fill these out. Once again, allow students to create their past work experiences, using the classified advertisements.

9. Distribute the Sample Cover Letters to students. Use the transparency of the sample and Writing a Cover Letter to discuss the different parts. Have students create their own cover letters to apply for specific jobs.

10. You may wish to conduct interviews for the jobs. If desired, a group of students or the entire class can be part of an interview team. If more than one student applies for a particular position, the interview team can vote to determine who gets it. Students who do not get the job would have to apply for other positions.

Follow-Up Activities

1. Have students draw pictures of themselves at work. Ask them to write descriptions of their professions.

2. Encourage students to talk to their parents about various professions.

3. Invite parents or community members to speak to your class about their professions.

4. Ask students to design their own letterheads to use for their cover letters.

Professions and Salaries

## Attorney (Lawyer) **Yearly Salary: $80,000**	## Physician (Doctor) **Yearly Salary: $90,000**
## Teacher **Yearly Salary: $28,000**	## Receptionist **Yearly Salary: $20,000**
## Secretary **Yearly Salary: $22,000**	## Bank Teller **Yearly Salary: $20,000**
## Salesperson **Yearly Salary: $20,000**	## Nurse **Yearly Salary: $40,000**

Professions and Salaries *(cont.)*

Construction Worker

Yearly Salary: $28,000

Accountant

Yearly Salary: $28,000

Counselor

Yearly Salary: $25,000

Computer Technician

Yearly Salary: $22,000

Dishwasher

Yearly Salary: $12,000

Waiter/Waitress

Yearly Salary: $12,000

Custodian (Janitor)

Yearly Salary: $14,000

Electrician

Yearly Salary: $30,000

Sample Job Application

APPLICATION FOR EMPLOYMENT

(Please print or type.)

Position Desired: _Sales Clerk_   Date Available to Begin: _Jan. 5_

Personal Information

Name: _Jones_   _Ray_   _Martin_
        (Last)   (First)   (Middle)

Address: _11211 Red Robin Drive_   _Austin_   _Texas_   _78753_
          (Street)   (City)   (State)   (Zip)

Telephone Number: _(512) 999-9999_   Social Security Number: _455-00-1234_

Previous Address: _181 Canyon Vista Road, Apt. #9   Austin, Texas 78759_

Education

	Name and Location	Years Attended	Date Graduated
High School	Anderson 935 Mesa Dr., Austin, Texas	1992–1996	May 1996
College	University of Texas Austin, Texas	1996–present	
Trade/Technical or Correspondence School	Electronic Technicians School Austin, Texas	1998–1999	August 1999

Work Experience (Start with the most recent.)

Company Name: _Food Mart_   Location: _Austin, Texas_

Job Title: _Cashier_   Dates Worked: from _5/95_ to _8/96_

Salary: _$5.00_ per _hour_   Reason for Leaving: _Starting college_

Company Name: _The Best Bookstore_   Location: _Austin, Texas_

Job Title: _Stocker_   Dates Worked: from _5/94_ to _8/94_

Salary: _$3.50_ per _hour_   Reason for Leaving: _Summer employment only_

References (Provide the names of three persons who are not relatives or employers.)

Name	Address and Phone Number	Occupation	Years Known
Dr. Paul Barrett	7680 Tallwood, Austin, TX 78757 (512) 999-1100	Physician	10
Mrs. Anna Peabody	4201 Merry, Austin, TX 78750 (512) 999-3232	Teacher	4
Mr. Zeb Osbourne	19101 Frierson, Austin, TX 78723 (512) 999-7693	Store Clerk	2

Date: _Dec. 15, 2000_   Signature: _Ray Martin Jones_

Job Application Form

APPLICATION FOR EMPLOYMENT

(Please print or type.)

Position Desired: _____   Date Available to Begin:_____

Personal Information

Name: _____
                (Last)   (First)   (Middle)

Address: _____
            (Street)   (City)   (State)   (Zip)

Telephone Number: _____   Social Security Number: _____

Previous Address: _____

Education

	Name and Location	Years Attended	Date Graduated
High School			
College			
Trade/Technical or Correspondence School			

Work Experience (Start with the most recent.)

Company Name: _____   Location: _____

Job Title:_____   Dates Worked: from_____ to _____

Salary:_____ per _____   Reason for Leaving: _____

Company Name: _____   Location: _____

Job Title: _____   Dates Worked: from_____ to _____

Salary:_____ per _____   Reason for Leaving: _____

References (Provide the names of three persons who are not relatives or employers.)

Name	Address and Phone Number	Occupation	Years Known

Date: _____   Signature: _____

Writing a Resumé

Explain to students that a resumé is a one- to two-page summary of their education, work experiences, and special training. The purpose of a resumé is to give an employer a quick look at what someone is capable of doing on the job.

To teach students about resumés, provide copies of the sample (page 22). Point out that there are many ways to organize a resumé and this format is only one of them. Use a transparency of the sample and the explanation shown below to discuss the parts of a resumé. Stress the importance of a resumé being short, clear, truthful, neat, and flawless. After presenting the sample, allow students to fill out their own resumés using the blank form (page 23). Have them write rough drafts. You may wish to have students help edit and proofread each other's rough drafts. Then ask them to write their final drafts.

Center the following information at the top:

Your Name
Your Address
Your Telephone Number

Education: List the college or colleges you attended, list any degrees that you earned, and note the dates of attendance or the year of graduation. Include your major field of study (if it is not obvious from the degree you earned) and any academic honors you received. The overall grade point average can also be shown.

Work Experience: List full-time employment, starting with your most recent job first. For each position, tell the title of the job, the dates of employment, and a brief description of the duties. Work back to the earliest job that is relevant to the work being sought. List part-time and temporary jobs only if there are few full-time positions that were held or if they are relevant to the work being sought. Point out that jobs lasting only a short time or periods of unemployment will need to be explained at an interview. Discuss with students how jobs in upper management will require an impressive job history, including years of experience and special skills.

Special Skills and Activities: This should be short but can include memberships in professional organizations, specialized skills such as foreign languages or computer programming, hobbies that are relevant to the work being sought, and a list of works that you have had published.

References: These are people an employer can contact who will tell how skillful you are or what kind of person you are. Most often the names of references are not given on a resumé. Instead write "Available upon request."

Students who finish before the rest of the class can draw pictures of themselves doing their jobs and write descriptions of their professions. Encourage them to do research to learn more about their professions so their descriptions are more complete. Arrange for guest speakers or allow students to call professional organizations and local businesses for information.

Sample Resumé

Susan Marie Dayton
4837 Foltan Avenue
Sherman Oaks, CA 91423
(818) 555-1000

Education

University of California at Los Angeles—Masters in Business Administration with high honors, 1998, GPA: 3.9

University of California at Los Angeles—Bachelor of Economics with high honors, 1996, GPA: 3.7

Honors: School of Business Honor Society, Freshman Honor Society

Work Experience

Bank of the United States, 9/96 – Present
(Los Angeles, California)

- Position: Financial Analyst
- Duties: Examine different financial plans; make recommendations to senior analyst; organize and make presentations regarding new investments

Town and Country Bank, 5/93 – 8/96
(Los Angeles, California)

- Position: Bank Teller
- Duties: Provided customer service; handled cash; maintained accurate records; managed customer accounts

Johnson & Johnson Accounting Firm, 6/92 – 8/92
(Los Angeles, California)

- Position: Clerk
- Duties: Filed client records; photocopied documents; organized professional library

Special Skills and Activities

Chairperson of the Committee for Better Banking Services
Member of the Association of Corporate Banking
Editor of *Technology in Banking,* quarterly report
Computer skills, including creating spreadsheets
Typing/Word processing (70 wpm)

References

Available upon request

My Resumé

_____

_____

_____

_____

Education

_____

_____

_____

Work Experience

_____

_____

_____

_____

_____

Special Skills and Activities

_____

_____

_____

References

_____

Writing a Cover Letter

Explain that a cover letter should express enthusiasm as well as show competence for a position, with the main goal of being chosen for an interview. A cover letter appears at the beginning of the materials to be included and is used to convince a prospective employer that the person writing it is the best candidate for the position.

Provide students with copies of the Sample Cover Letter (page 25). Use a transparency of the sample and the explanation shown below to discuss the parts of a cover letter. Stress the importance of a cover letter being short, clear, truthful, neat, and flawless. After presenting the sample, allow students to pick jobs from the classified ads. Ask them to write rough drafts of their own cover letters to apply for these positions. You may wish to have students help edit and proofread each other's rough drafts. Then ask them to write their final drafts.

(Print or type all the information flush left.)

Your Street Address
Your City, State, Zip
Date

Name of Contact Person*
Position/Title of Contact Person*
Company Name*
Company Address*
*(*This information can be made up if it is not stated in the classified ad.)*

Greeting: Use "Dear Mr.____" or "Dear Ms.____" if the contact's name is known. Otherwise, use "Dear Personnel Manager." Write a colon at the end.

(Do not indent the paragraphs. Skip a space between paragraphs.)

Paragraph 1: This paragraph is the introduction and states your purpose for writing the letter. Tell the position you are applying for and where it was advertised.

Paragraph 2: This paragraph is the body of the letter and is used to summarize your qualifications, including education and work experiences. Refer to the resumé for the specific details. Focus on selling yourself, explaining how your special talents are perfect for the job.

Paragraph 3: This paragraph is the conclusion and should focus on encouraging the employer to contact you for an interview. You should restate your desire to work for the company, specify where you can be contacted, and tell when you will be available for an interview.

Closing: Use "Sincerely" or "Sincerely yours" followed by a comma.

Signature: Use cursive handwriting to sign your whole name.

Printed Name: Print or type your whole name under your signature so it is easy to read.

Enclosure: Resumé *(Point out that you are including a resumé with your cover letter.)*

Sample Cover Letter

9217 Greystone Drive
Topeka, Kansas 66614
April 27, 1999

David Arnold
Director of Special Investigations
Davenport Detective Agency
1127 Frier Drive
Los Angeles, CA 90210

Dear Mr. Arnold:

I am writing in response to your advertisement in the *Topeka Times*. I would like to apply for the position of special investigator.

As my enclosed resumé indicates, I have had extensive experience in the detective and insurance businesses, making me a prime candidate for this position. I have a degree in law enforcement and worked as a police officer in Topeka for five years. I received three decorations during that time. For the next two years, I worked as a police detective. After leaving the police force, I became a well-respected private investigator. I was hired a number of times to do insurance investigations for Golden Eagle, Incorporated. Over the years, I have earned a reputation for being hardworking and maintaining a positive attitude. I feel that all of these experiences have given me excellent interpersonal and leadership skills, allowing me to be a successful investigator for your company.

I would like to discuss this matter further with you. I can be reached at (913) 555-2121. I am available for interviews at any time that is convenient for you.

Sincerely,

Susan Marie Dayton
Susan Marie Dayton

Enclosure: Resumé

Checking Accounts

Objectives

Students will write checks and fill out deposit slips. They will be able to record withdrawals and deposits in check registers. Students will review adding and subtracting with decimals.

Suggested Vocabulary

financial institution/bank/credit union, checking account, savings account, withdrawal, deposit, balance, interest, endorse, insufficient funds, overdrawn, bounce a check, credits, debits, monthly bank statement

Materials

- Sample Check and Deposit Slip (page 27), transparency and one copy per student
- Check and Deposit Slip Forms (page 28), several copies per student
- Keeping a Check Register (page 29), transparency and one copy per student

Optional:

- Understanding Decimals (page 83)
- Adding and Subtracting Decimals (page 84)
- Answer Key (page 93)

Procedure

1. Ask students to brainstorm a list of things for which they might pay using checks. Then have them create a list of reasons they might need to make deposits.

2. Distribute copies of the Sample Check and Deposit Slip to students. Show the transparency, and discuss how to complete these forms.

3. Give the blank Check and Deposit Slip Forms to students. Ask students to practice writing checks to local businesses. Have them practice writing deposit slips for different amounts.

4. Discuss the purpose of keeping a check register. Provide copies of the register to students. Use the transparency to talk about the sample. Ask volunteers to help you make withdrawals and deposits on the blank register at the bottom of the transparency.

5. Encourage students to make their own registers for some imaginary deposits and withdrawals.

Follow-Up Activities

1. Take a field trip to a local bank, or invite a bank employee to speak to your class.

2. Show students a variety of check designs. Then have them create their own designs.

3. Have students compare/contrast different local banks and what they offer customers.

Sample Check and Deposit Slip

Carefully examine the sample check and deposit slip shown below. Practice writing checks and deposit slips using the forms on page 28.

SAMPLE CHECK

Amber West
1309 1st Street
Huntington Village, CA 92000
(700) 801-9901

331

Mar. 1 19 *99*   98–7170/3341

Pay to the order of _*Foods Plus*_ $ *37.12*

*Thirty—seven and* 12/100 Dollars

Bank of Huntington Beach
Huntington Beach, California 92647

For _*groceries*_   *Amber West*

⑆341971401⑆ 3940561278⑈

SAMPLE DEPOSIT SLIP

Deposit Ticket

Name _*Amber West*_

Date _*Mar. 10,*_ 19 *99*

*Amber West*
Sign here if cash received from deposit.

Bank of Huntington Beach
Huntington Beach, California 92647

⑆341971401⑆ 3940561278⑈

Currency		
Coin		
Checks — List Checks Singly	15	25
	150	00
	94	10
Total From Other Side	65	83
Total	325	18
Less Cash Received	100	00
Net Deposit	225	18

98–7170/3341

Check and Deposit Slip Forms

Review the sample check and deposit slip on page 27. Practice writing checks and deposit slips using the forms shown below.

CHECK

Amber West
1309 1st Street
Huntington Village, CA 92000
(700) 801-9901

331

_____19_____   98–7170/3341

Pay to the
order of _____ $_____

_____   Dollars

Bank of Huntington Beach
Huntington Beach, California 92647

For _____   _____

⑆341971401⑆ 3940561278⑈

DEPOSIT SLIP

Deposit Ticket

Name_____

Date_____ 19_____

_____
Sign here if cash received from deposit.

Bank of Huntington Beach
Huntington Beach, California 92647

Currency		
Coin		
Checks List Checks Singly		
Total From Other Side		
Total		
Less Cash Received		
Net Deposit		

98–7170/3341

⑆341971401⑆ 3940561278⑈

Keeping a Check Register

Review the sample check register shown below.

		Record all transactions that affect your account.						
Number	Date	Description of Transaction	Payment or Debit (−)	Tax Item (✓)	Fee— if any	Deposit or Credit (+)	Balance	612 20
261	1/8	Pets, Pets, Pets	24 99					− 24 99
		dog food						587 21
262	1/8	City of Chicago	120 75					− 120 75
		electric bill						466 46
263	1/9	Food World	68 16					− 68 16
		groceries						398 30
264	1/10	VOID	———					———
								398 30
——	1/10	Deposit				200 00		+ 200 00
		birthday checks						598 30

Now practice completing the following check register to keep a record of deposits/credits and payments/debits.

		Record all transactions that affect your account.						
Number	Date	Description of Transaction	Payment or Debit (−)	Tax Item (✓)	Fee— if any	Deposit or Credit (+)	Balance	

Family Budgets

Objectives

Students will predict and estimate their monthly and yearly family budgets for food, housing, transportation, other expenses, unexpected events, and total expenses. Then they will compare their predictions with their estimates.

Suggested Vocabulary

See pages 31, 47, 53, 59, 67, and 74.

Materials

- Part 1: Food (pages 31–46)
- Part 2: Housing (pages 47–52)
- Part 3: Transportation (pages 53–58)
- Part 4: Other Expenses (pages 59–66)
- Part 5: Unexpected Events (pages 67–73)
- Part 6: Total Expenses (pages 74–80)
- Family Profile Record Sheets (page 10) from previous simulations.

Procedure

1. Ask students to brainstorm a list of possible expenses a family might have. Discuss the purpose and the importance of having a budget. Explain that a budget is the amount of money a family thinks it will need to cover its expenses.

2. Before beginning Simulation # 4: Part 1, have students predict their monthly and yearly budgets for food, housing, transportation, other expenses, unexpected events, and total expenses. Tell them to write this information on their Family Profile Record Sheet (page 10).

3. First, have students use prior knowledge to predict their monthly costs. Then tell them to determine their yearly costs by multiplying each monthly value by 12 (for the number of months).

4. Have students determine the monthly and yearly total expenses. Explain that the monthly total is determined by adding together all of the monthly values for each category. The yearly total can be determined by adding together all of the yearly values for each category or by taking the total monthly amount and multiplying it by 12. Allow students to use calculators if needed.

5. Discuss how budgets may vary, depending on family size and needs.

Follow-Up Activities

1. Encourage students to keep a record of their real-life families' basic expenses for one or two months. You may wish to limit this activity to one category such as food, transportation, or housing.

2. Invite a financial planner to speak to your class.

Food

Objectives

Students will create sample food menus to plan breakfast, lunch, and dinner meals. Then they will estimate their food expenses for one day, one week, one month, and one year.

Suggested Vocabulary

brand name, generic, nutritious, fresh produce, dairy products, poultry, frozen foods, entrees, beverages, bakery, label, ingredients, calories, preservatives, additives, artificial flavors/colors, lowfat, sugar free, price per unit, bulk, coupon, clerk/cashier, register, customer, restaurant, check/bill, fast foods, tip, à la carte

Materials

- Food Pyramid (page 34), one copy per student
- Food Price List (pages 35–38), one copy per student or for pairs of students
- Sample Food Menu—Option 1 (page 39), one copy per student if using Option 1
- My Food Menu—Option 1 (page 40), 21 copies per student if using Option 1
- Food Menu Summary—Option 1 (pages 41–43), one copy per student if using Option 1
- Sample Food Menu—Option 2 (page 44), one copy per student if using Option 2
- My Food Menu—Option 2 (page 45), three copies per student if using Option 2
- Food Menu Summary—Option 2 (page 46), one copy per student if using Option
- Family Profile Record Sheets (page 10) from previous simulations.

Optional:

- Basic Operations with Whole Numbers (pages 81 and 82)
- Understanding Decimals (page 83)
- Adding and Subtracting Decimals (page 84)
- Understanding Fractions and Mixed Numbers (page 85)
- Multiplying Fractions and Mixed Numbers (page 86)
- Understanding Percentages (page 87)
- Finding Percentages (page 88)
- Answer Key (page 93)
- Measurement Conversion Chart (page 96)

Procedure

1. Be sure students know how to multiply fractions and convert fractions to decimals for the food menu activities.

2. For capable math students, use Option 1 (page 32, pages 39–43). For less capable math students, use Option 2 (page 33, pages 44–46). Provide students with copies of the Food Pyramid (page 34) whether they are using Option 1 or 2. Have students do research at local grocery stores to determine food prices or provide copies of the Food Price List (pages 35–38).

3. After students complete Option 1 or 2, be sure they record their estimated monthly and yearly expenses for food on their Family Profile Record Sheets (page 10).

Food *(cont.)*

Procedure *(cont.)*

Option 1:

1. Emphasize the importance of a healthful and balanced diet. Discuss the food pyramid. To simplify the process of creating healthful meals, instruct students to have each meal contain one item from each of the four major food groups. For students wishing to be vegetarians or requesting special dietary needs, direct them to alternative protein sources or, under special circumstances, allow substitutions with items from other food groups. As they plan their meals, encourage students to dine out or get takeout no more than twice per week or eat at a friend's house (for free) no more than one meal per week.

2. Allow students to be creative as they plan their menus. Some of them may want to grow their own fruits and vegetables. This is permissible as long as the prices for seeds, soil, pots, water, etc., are researched and divided by 52 to get an approximate weekly cost. Then this value should be added to the other weekly costs on the food menu. You may wish to have students who choose to grow their own fruits and vegetables also calculate their meals based on the price list and then compare the two results.

3. Since Option 1 can be time consuming and challenging, you may wish to have students use calculators or work with partners so they can help each other. Discuss the format for the food menu, using the sample. Point out how to multiply the fraction used by the dollar amount and how to convert the answer to decimal form on the Food Menu Summary. Provide guided practice and sample calculations based on various family sizes.

4. The key point of the calculations in Option 1 is to keep track of the values represented by the numerator and denominator in each fraction. The numerator represents how many servings are being used for a specific meal, while the denominator represents how many total servings are available when purchasing the whole food item.

5. Point out to students that when the number of total servings available for a certain food item is not enough to feed the entire family, more than one container/package needs to be purchased. For example, a can of tuna that costs $0.75 serves two people. For a family of four, you would need to buy two cans of tuna. Therefore, the price for tuna to feed a family of four would be $1.50.

6. Encourage students to write comments on page 43 about their completed Food Menu Summary. Students can then discuss and compare their findings.

7. For your more advanced students, encourage the use of actual store coupons to reduce food costs. The coupons have to be for items found on the price list. To simplify the calculations, have students deduct the total value of the coupons from the weekly total rather than trying to determine what fraction of the coupon they are using for each meal.

Food *(cont.)*

Procedure *(cont.)*

Option 2:

1. This method of completing the food menu is shorter and simpler than Option 1. Instead of planning meals for seven days, students plan meals for only one day and then multiply the day's total by seven to get a weekly cost. If you prefer, have students determine the meals for seven days, using the Option 2 format.

2. Instruction can be individualized, having some groups of students work with Option 1 while other groups work with Option 2. The monthly and yearly costs determined in the Food Menu Summary would be calculated in the same way for either option. Students will multiply the weekly total by 4 to get the monthly total and then multiply the monthly total by 12 to get the yearly total.

3. Have students prepare their food menus by determining the amount of food needed to provide three balanced meals per day for each family member. You may wish to have students work independently or with partners.

4. Using the Food Price List, students figure out the cost for each family meal. These calculations will include adding, multiplying, and dividing decimals; multiplying fractions; and determining percentages. It may be helpful for some students to use calculators.

5. When the summaries are completed, encourage students to write comments on page 46 and to share their findings.

6. For your more advanced students, encourage the use of actual store coupons to reduce food costs. The coupons have to be for items found on the price list. To simplify the calculations, have students deduct the total value of the coupons from the weekly total rather than trying to determine what fraction of the coupon they are using for each meal.

Follow-Up Activities

1. Discuss how some families save money on food costs by using coupons, buying in bulk, buying day-old bakery items, using leftovers, eating at home rather than dining out, etc.

2. Invite a nutritionist or dietitian to speak to your class.

3. Have students prepare some simple recipes.

4. Tell each student to bring a copy of his/her favorite recipe from home. Use the recipes to create a class cookbook. Reproduce the cookbook, and distribute it to students.

5. Encourage students to learn about foods from different cultures.

6. For more advanced students, have them research food prices and food shortages in countries around the world.

Food Pyramid

A Guide to Daily Food Choices

In April of 1992, the United States Department of Agriculture created a food pyramid which shows the kinds of foods and numbers of servings needed each day in order to stay healthy.

Key
● Fat (naturally occurring and added)
▼ Sugars (added)

These symbols show that fat and added sugars come mostly from fats, oils, and sweets but can be part of or added to foods from the food groups as well.

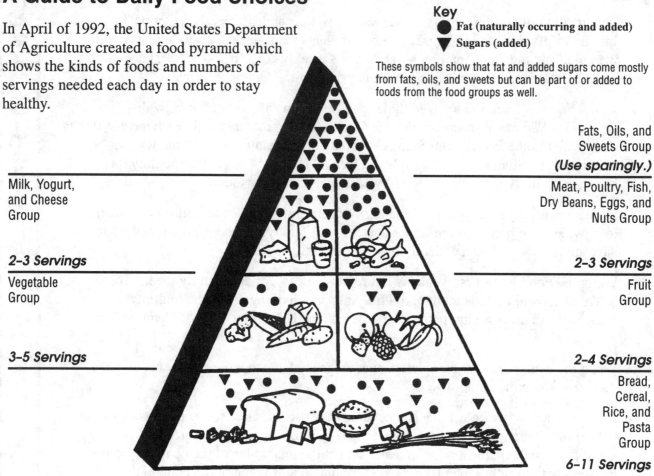

Fats, Oils, and Sweets Group
(Use sparingly.)

Milk, Yogurt, and Cheese Group

Meat, Poultry, Fish, Dry Beans, Eggs, and Nuts Group

2–3 Servings

2–3 Servings

Vegetable Group

Fruit Group

3–5 Servings

2–4 Servings

Bread, Cereal, Rice, and Pasta Group

6–11 Servings

Serving Sizes

Bread, Cereal, Rice, and Pasta

1 slice of bread
½ cup of cooked rice or pasta
½ cup of cooked cereal
1 ounce of ready-to-eat cereal

Vegetables

½ cup of chopped raw or cooked vegetables
1 cup of leafy raw vegetables

Fruits

1 piece of fruit or melon wedge
¾ cup of juice
½ cup of canned fruit
¼ cup of dried fruit

Milk, Yogurt, and Cheese

1 cup of milk or yogurt
1 ½ to 2 ounces of cheese

Meat, Poultry, Fish, Dry Beans, Eggs, and Nuts

2 ½ to 3 ounces of cooked lean meat, poultry, or fish
Count ½ cup of cooked beans, or 1 egg, or 2 tablespoons of peanut butter as 1 ounce of lean meat (about ⅓ serving).

Fats, Oils, and Sweets

LIMIT CALORIES FROM THESE, especially if you need to lose weight.

The amount you eat may be more than one serving. For example, a dinner portion of spaghetti would count as two or three servings of pasta.

Food Price List

Meat, Poultry, Fish, Dry Beans, Eggs, and Nuts Group

Chicken
$1.00 per piece (single serving)
$0.69 per pound
$5.00 whole chicken (serves 4)

Turkey
$3.00 per pound (serves 4)

Lamb
$1.00 per chop (single serving)
$4.00 per package of 6 chops

Steak
$2.00 per piece (single serving)
$6.00 per package of 4 pieces

Hot Dogs
$2.00 per package of 6 hot dogs

Veal
$1.00 per piece (single serving)
$4.00 per package of 5 pieces

Tuna
$0.75 per can (serves 2)

Peanut Butter
$2.00 per jar (serves 12)

Hamburger
$2.00 per pound (serves 4)

Fish
$1.00 per piece (single serving)
$3.00 per package of 4 pieces

Food Price List *(cont.)*

Meat, Poultry, Fish, Dry Beans, Eggs, and Nuts Group *(cont.)*

Roast Beef or Ham

$6.00 per pound (serves 4)

Beans

$0.50 per serving

Bacon

$3.00 per pound (20 slices)

Nuts

$2.00 per serving

Eggs

$2.00 per dozen

Tofu

$1.00 per serving

Milk, Yogurt, and Cheese Group

Milk
$1.00 per quart (serves 2)
$2.00 per half gallon (serves 4)
$3.00 per gallon (serves 8)

Cream Cheese
$2.00 per package
(serves 6)

Yogurt
$0.75 per 8 oz. container
(serves 1)

Cottage Cheese
$1.50 per 16 oz. container
(serves 2)

American Cheese
$3.00 per package of 16 slices

Other Cheeses
$2.00 per pound (serves 6)

Food Price List *(cont.)*

Fruit Group

Fresh Fruit (all)

$1.00 per pound (serves 4)

Canned Fruit

$0.75 per can (serves 2)

Fruit Juice

$1.50 per quart (serves 2)

$2.75 per half gallon (serves 4)

$4.00 per gallon (serves 8)

Vegetable Group

Fresh Vegetables (all)

$1.00 per pound (serves 4)

Canned Vegetables

$0.75 per can (serves 2)

Vegetable Juice

$1.50 per quart (serves 2)

Bread, Cereal, Rice, and Pasta Group

Bread

$2.00 per loaf (20 slices)

$1.50 per package of 12 buns

Bagels

$2.00 per bag of 6 bagels

Pasta

$2.00 per package (serves 4)

Cereal

$3.00 per box (serves 8)

Rice

$1.00 per bag (serves 6)

Flour

$1.00 per pound (20 servings)

Food Price List *(cont.)*

Fats, Oils, and Sweets and Others

Sugar

$2.00 per pound (20 servings)

Cakes/Pies

$5.00 each (serves 8)

Soda

$2.00 per six-pack

Butter/Margarine

$1.75 per four-stick pack
(48 servings)

Pizza

$7.00 small (single serving)

$10.00 medium (serves 2)

$15.00 large (serves 4)

Soup

$1.00 per can (serves 2)

Candy

$0.50 per bar

$3.00 per pound (serves 4–6)

Chips

$2.00 per bag (10 servings)

Ice Cream

$3.00 per gallon (serves 12)

Condiments

(catsup, mustard, salad dressing,
etc.)

$2.00 each (20 servings)

Frozen Dinners

$1.50 small (single serving)

$3.00 large (single serving)

Spices

$3.00 per jar (20 servings)

Dining Out

Better Restaurant

$20.00 per adult, $10.00
per child

Fast Food Restaurant

$6.00 per adult, $3.00 per child

38

Sample Food Menu—Option 1

Family: 2 adults, 3 children

Meal: Breakfast

Day: (1)   2   3   4   5   6   7   (Circle one.)

Food Item	(Servings per Adult x Number of Adults) + (Servings per Child x Number of Children) =	Total Servings
Eggs	(2 x 2) + (1 x 3) = 4 + 3 =	7
Toast	(2 x 2) + (1 x 3) = 4 + 3 =	7
Fresh Fruit	(1 x 2) + (1 x 3) = 2 + 3 =	5
Milk	(1 x 2) + (1 x 3) = 2 + 3 =	5

Food Item	Calculation of Cost	Cost for No. of Servings Needed (Rounded to Nearest Cent)
Eggs	\$2.00 for 12—only 7 needed $7/12$ of \$2.00 = $7/12$ x $2/1$ = $14/12$ $14/12$ = 14 ÷ 12 = \$1.17	\$1.17
Toast	\$2.00 for 20 slices—only 7 needed $7/20$ of \$2.00 = $7/20$ x $2/1$ = $14/20$ $14/20$ = 14 ÷ 20 = \$0.70	\$0.70
Fresh Fruit	\$1.00 for 1 pound (4 servings)—not enough \$2.00 for 2 pounds (8 servings)—5 are needed $5/8$ of \$2.00 = $5/8$ x $2/1$ = $10/8$ $10/8$ = 10 ÷ 8 = \$1.25	\$1.25
Milk	\$3.00 for 1 gallon (8 servings)—only 5 are needed $5/8$ of \$3.00 = $5/8$ x $3/1$ = $15/8$ $15/8$ = 15 ÷ 8 = \$1.88	\$1.88
	TOTAL	**\$5.00**

My Food Menu—Option 1

Family:

Meal:

Day: 1   2   3   4   5   6   7   (Circle one.)

Food Item	(Servings per Adult x Number of Adults) + (Servings per Child x Number of Children) =	Total Servings

Food Item	Calculation of Cost	Cost for No. of Servings Needed (Rounded to Nearest Cent)
	TOTAL	

Food Menu Summary—Option 1

After you have created your food menu on page 40, use the information to determine the cost of your family's meals. Complete pages 41– 43 to estimate your weekly, monthly, and yearly food expenses.

Day 1

Cost of Breakfast = _____

Cost of Lunch = _____

Cost of Dinner = _____

Total—Cost for Day 1 (add amounts listed above) = _____

Day 2

Cost of Breakfast = _____

Cost of Lunch = _____

Cost of Dinner = _____

Total—Cost for Day 2 (add amounts listed above) = _____

Day 3

Cost of Breakfast = _____

Cost of Lunch = _____

Cost of Dinner = _____

Total—Cost for Day 3 (add amounts listed above) = _____

Food Menu Summary—Option 1 *(cont.)*

Day 4

Cost of Breakfast = _____

Cost of Lunch = _____

Cost of Dinner = _____

Total—Cost for Day 4 (add amounts listed above) = _____

Day 5

Cost of Breakfast = _____

Cost of Lunch = _____

Cost of Dinner = _____

Total—Cost for Day 5 (add amounts listed above) = _____

Day 6

Cost of Breakfast = _____

Cost of Lunch = _____

Cost of Dinner = _____

Total—Cost for Day 6 (add amounts listed above) = _____

Food Menu Summary—Option 1 *(cont.)*

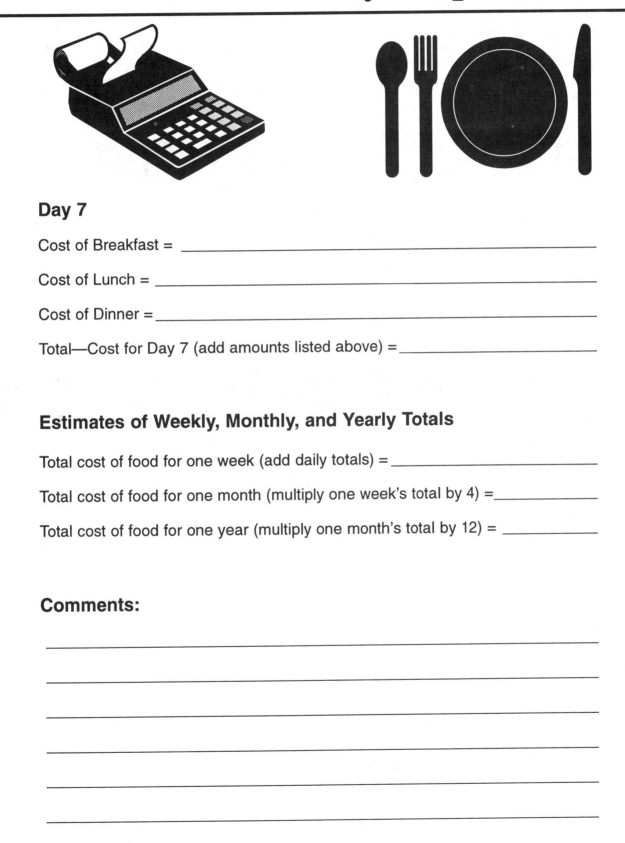

Day 7

Cost of Breakfast = _____

Cost of Lunch = _____

Cost of Dinner = _____

Total—Cost for Day 7 (add amounts listed above) = _____

Estimates of Weekly, Monthly, and Yearly Totals

Total cost of food for one week (add daily totals) = _____

Total cost of food for one month (multiply one week's total by 4) =_____

Total cost of food for one year (multiply one month's total by 12) = _____

Comments:

_____

_____

_____

_____

_____

_____

Sample Food Menu—Option 2

Family: 2 adults, 3 children

Meal: Breakfast

Food Item	(Servings per Adult x Number of Adults) + (Servings per Child x Number of Children) =	Total Servings
Eggs	(2 x 2) + (1 x 3) = 4 + 3 =	7
Toast	(2 x 2) + (1 x 3) = 4 + 3 =	7
Fresh Fruit	(1 x 2) + (1 x 3) = 2 + 3 =	5
Milk	(1 x 2) + (1 x 3) = 2 + 3 =	5

Food Item	Calculation of Cost	Cost for No. of Servings Needed (Rounded to Nearest Cent)
Eggs	($2.00 ÷ 12 eggs) x 7 = $1.17	$1.17
Toast	($2.00 ÷ 20 slices) x 7 = $0.70	$0.70
Fresh Fruit	($2.00 ÷ 8 servings) x 5 = $1.25	$1.25
Milk	($3.00 ÷ 8 servings) x 5 = $1.88	$1.88
	TOTAL	**$5.00**

My Food Menu—Option 2

Family: _____

Meal: _____

Food Item	(Servings per Adult x Number of Adults) + (Servings per Child x Number of Children) =	Total Servings

Food Item	Calculation of Cost	Cost for No. of Servings Needed (Rounded to Nearest Cent)
	TOTAL	

Food Menu Summary—Option 2

Estimate of Cost for Daily Total

(The cost of each meal should be calculated using My Food Menu—Option 2 on page 45.)

Cost of breakfast =_____

Cost of lunch =_____

Cost of dinner = _____

Total cost of food for one day (add totals from all three meals) = _____

Estimate of Weekly Total

Total cost of food for one week (multiply one day's total by 7) = _____

Estimate of Monthly Total

Total cost of food for one month (multiply one week's total by 4) = _____

Estimate of Yearly Total

Total cost of food for one year (multiply one month's total by 12) = _____

Comments:

_____

_____

Housing

Objectives

Students will estimate their housing expenses for one month and one year. They will draw floor plans of their homes, showing the dimensions and area of each room.

Suggested Vocabulary

residence, residential area, apartment, condominium, utilities, insurance, address, payment, manager, resident, tenant, owner, lease/rent

Materials

- Housing Price List (page 50), one per student or pair of students
- My Housing (page 51), one per student
- My Floor Plan (page 52), one per student
- Family Profile Record Sheets (page 10) from previous simulations

Optional:

- Basic Operations with Whole Numbers (pages 81 and 82)
- Understanding Perimeter (page 89)
- Finding Perimeters (page 90)
- Understanding Area (page 91)
- Finding Areas (page 92)
- Answer Key (page 93)
- Centimeter Rulers (page 94)
- Inch Rulers (page 95)
- Measurement Conversion Chart (page 96)

Procedure

1. Discuss with students the advantages and disadvantages of owning versus renting property.

2. Have students use the Housing Price List to decide whether they will rent or own houses or condominiums, rent apartments, or build houses. Their choices should take into account both their personal preferences and their incomes. If they wish to buy property, they need to have enough money in their savings or checking accounts to pay 10% of the total purchase price up front. If they do not have enough money for this down payment, they may only rent property.

3. Remind students that they need to pick housing that is big enough for their family to live in while keeping the cost affordable. Tell them that the number of children per bedroom is limited to two.

4. You may wish to have students look in the newspaper for lower housing prices in your community. If they can show proof of a lower price for the same sized house, condominium, or apartment, allow them to use these values in their calculations.

Procedure (cont.)

5. Discuss the difference between area and perimeter. Review how to calculate area. Using the total size and room limitation provided in the price list, have students determine which rooms are in their homes and the area for each room. Have students record the information on My Housing.

6. Direct students to the portion of My Housing in which they list the names and sizes of their rooms. The names of the rooms (bedroom, bathroom, living room, dining room, kitchen, etc.) are given in the descriptions on the Housing Price List. Instruct students to mark which bedrooms are for children and which one is for the adult(s). Students with small families who may not need extra bedrooms may substitute the bedroom area for offices, guest bedrooms, or playrooms.

7. Challenge students to have the total area of the rooms come as close to the size limit as possible without going over it. The total area of the rooms should not be more than 20 square feet (6 square meters) under the size limitation. If the total area is too small or students go over the size limit, they need to recalculate the area of each room.

8. If students have difficulty choosing an appropriate length and width for a particular room in their homes, have them approximate these measurements by using their feet as measuring tools. Use the classroom size as a guide to demonstrate what the dimensions are by counting the steps needed to walk the length and width of the room. Show how to multiply the length times the width to get a rough estimate of the classroom's area.

9. Have students discuss whether the area of the classroom would be an appropriate size for a living room, dining room, kitchen, bedroom, bathroom, or any other room of a home. If the classroom measures approximately 20 by 25 feet (6 by 7.5 meters), students might feel that this is too big for a bedroom. They could decide that a better size for a bedroom would be 15 by 20 feet (4.5 by 6 meters). If students are still unsure the area chosen is appropriate, encourage them to walk out the approximate length and width in the classroom to see what the desired area would look like.

10. Encourage students to be creative by adding a yard, front or back; garden area; patio; or balcony. Explain that these are not included as "living space" when determining square footage. Point out that the sizes of these extras should be appropriate to the sizes of the housing.

11. Have students calculate the total cost of utilities and other household expenses. (See the Housing Expenses per Month chart on page 49.) Tell them to record these expenses on My Housing. Then have them put the total estimated cost of housing on their Family Profile Record Sheets (page 10).

Housing *(cont.)*

Procedure (cont.)

Housing Expenses per Month

Utilities		Household Supplies	
Electric, Gas, Trash		(cleansers, paper towels, mops, etc.)	
House	$200	Per household	$100
Condominium	$110		
Apartment	$ 75	**Homeowners' Fees**	
Water		(for water, grounds upkeep, repairs, etc.)	
House	$100	Per household	$ 50
Condominium	Paid in Homeowner's Fee	Condominium	$150
Apartment	Paid in Rent		
Telephone		**Lawn/Garden Upkeep**	
Per household	$ 50	House	$ 30

12. Show students a variety of floor plans. Provide transparent rulers (standard or metric measurement) or make transparency rulers (see pages 94 and 95). Challenge students to draw floor plans of their homes that are to scale. Tell them to scale down the dimensions they have chosen for each room. For example, 1 inch could represent 5 feet. Be sure students include measurement keys on their floor plans to show what scale was used.

Follow-Up Activities

1. Invite an architect, interior designer, builder, and/or construction worker to speak to your class. You may wish to have students prepare a list of questions ahead of time.

2. Arrange ahead of time to take your class to a construction site. Allow them to view houses/apartments that are at different stages of completion.

3. You may wish to have students determine the cost for furniture, using advertisements from a newspaper. Then have students choose how to purchase the furniture. They can pay the entire amount from their checking or savings accounts if they have sufficient funds, or they can put down a 10% deposit and make monthly payments for the remaining amount.

4. More advanced learners may enjoy using their floor plans to make models of their homes. Allow them to be creative with their models. They may wish to make furniture to place in each room of the model.

Housing Price List

TO RENT

House

Cost: $2,000 per month

Floor Plan: 3 bedrooms, 2 bathrooms, kitchen, living room, and dining room

Size Limit: 2,400 square feet

Condominium

Cost: $1,200 per month

Floor Plan: 2 bedrooms, 1 bathroom, kitchen, and combined living/dining room

Size Limit: 1,000 square feet

Apartment

Cost: $600 per month

Floor Plan: 1 bedroom, 1 bathroom, combined kitchen/living room

Size Limit: 600 square feet

TO PURCHASE

House

Cost: $130,000 paid in full or 10% down and $650 per month

Floor Plan: 3 bedrooms, 2 bathrooms, living room, dining room, kitchen, and 1 extra room

Size Limit: 3,000 square feet

Custom Built House

Cost: $100 per square foot, neighborhood with moderately-priced homes

Cost: $200 per square foot, neighborhood with expensive homes

Floor Plan and Size: Determined by builder

Condominium

Cost: $95,000 paid in full or 10% down and $425 per month

Floor Plan: 2 bedrooms, 1 bathroom, kitchen, and combined living/dining room

Size Limit: 1,000 square feet

My Housing

Information About My Home

Type of housing (apartment, condominium, house)? _____

Rent or own? _____

Size limit? _____

Number of rooms allowed? _____

Name of Room (bedroom, bathroom, kitchen, living room, dining room, extra room)	Area in Square Feet (Meters) (area = length x width)
1.	
2.	
3.	
4.	
5.	
6.	
7.	
8.	
9.	
10.	
TOTAL AREA	

- Is the total area more than the size limit? _____ If your answer is *yes*, you need to recalculate the area of each room until your total area is less than the size limit.

- If you subtract the total area from the size limit, do you have more than 20 square feet (6 square meters) left? _____ If your answer is *yes*, you need to recalculate the area for each room until you have fewer than 20 square feet (6 square meters) left.

Money Spent on My Home	Per Month	Per Year
Rent or mortgage		
Utility bills (phone, gas, electricity, water, trash)		
Household supplies (cleansers, paper towels, mops, etc.)		
Other household expenses (grounds, homeowners' fees, furniture, etc.)		
TOTAL AMOUNT		

Now use page 52 to draw a floor plan of your home.

My Floor Plan

Transportation

Objectives

Students will determine the monthly and yearly costs of owning one or two cars. They will draw pictures of their cars.

Suggested Vocabulary

automobile, Department of Motor Vehicles, driver's license, license plate, loan, payment, insurance premiums, mechanic, repair, alarm, public transportation, traffic accident, inspection, unleaded gasoline, tune-up, gas station/service station

Materials

- Transportation Price List (page 56), one per student or pair of students
- Transportation Summary (page 57), one per student
- My Car (page 58), one per student
- Family Profile Record Sheets (page 10) from previous simulations

Optional:

- Basic Operations with Whole Numbers (pages 81 and 82)
- Answer Key (page 93)
- Measurement Conversion Chart (page 96)

Procedure

1. Divide the class into cooperative learning groups. Ask the groups to discuss the pros and cons of buying a new car and a used car. Have each group create a chart that shows this information.

2. Ask students to review the costs for purchasing and insuring different kinds of cars that are given on the Transportation Price List so they can decide whether they can afford to buy new or used cars for their families.

3. Other forms of transportation, such as city buses, commuter trains, motorcycles, and bicycles, are not considered in this cost analysis but can be discussed with students.

4. You may wish to have students look through the automobile sales in the classified advertisements of the newspaper. If they can find new or used car prices that are better than the approximate ones provided on the price list, allow them to use these values in their calculations.

5. Discuss the three groups of new cars shown on the Transportation Price List. Have students give examples of economy, mid-price, and luxury cars. Then talk about the conditions of used cars. Ask volunteers to describe what they think fair, good, and excellent conditions mean.

6. Ask students why they think automobile insurance is important. Invite an automobile insurance agent to speak to your class. Encourage students to ask the agent questions about what services the agency provides and what the different costs are for new and used cars.

Transportation *(cont.)*

Procedure (cont.)

7. Have students record their car insurance costs on their Transportation Summary forms.

8. Students should decide whether they will make monthly payments or pay for their cars in full, using money in their savings or checking accounts. If students purchase their cars, they will not have any monthly car payments. Therefore, the value recorded on their Transportation Summaries will be zero. Point out that they will still have to pay for car insurance and gasoline.

9. Write the following gas prices on the chalkboard or have students determine what the price of each grade of gas is in your community and write that information on the chalkboard.

Gas Prices

$1.14 per gallon (3.79 liters) of regular unleaded

$1.20 per gallon (3.79 liters) of mid-grade unleaded

$1.35 per gallon (3.79 liters) of premium unleaded

---

10. Tell students to calculate the amount of money they spend on gas if they use 18 gallons per week. After students have calculated the weekly cost of gasoline, have them multiply that value by four to get the monthly cost. Remind them to record the monthly value on their Transportation Summary forms.

11. Have students calculate the cost, insurance, and gasoline for a second car if their families need vehicles for their spouses.

12. Encourage students to draw pictures of their family cars.

13. As an alternative to drawing pictures, you may wish to have students make collages, using car pictures from newspapers and magazines.

14. Remind each student to write the total estimated transportation expenses on his or her Family Profile Recording Sheet (page 10). Be sure students write monthly and yearly estimates. Point out that yearly estimates can be obtained by multiplying the monthly estimates by 12.

Transportation *(cont.)*

Follow-Up Activities

1. If possible, arrange to take students to a car manufacturing plant. As an alternative, you could arrange a field trip to an automotive repair or body shop.

2. Have students do research to learn about vehicles that run on alternative fuel sources such as natural gas, electricity, solar power, human power, and hydrogen.

3. You may wish to have students investigate to learn what the different advantages and disadvantages of purchasing and leasing automobiles might be.

4. Obtain automobile accident report forms from an insurance company and the local police department. Reproduce these forms. Give students the experience of filling out these forms.

5. Arrange to take students to traffic court. Tell them to take notes during the hearings. Discuss students' observations of the process. Ask them what they think works well and what they think can be done to make improvements.

6. Have students do research to determine how gas is obtained, how it is processed, how it gets to the pumps, and how the price is established.

7. Provide students with outline maps of the world on which they can write. Have them do research to learn where the world's largest oil deposits are found. Ask them to color and label these locations on their maps.

8. Have students keep a log of how much gas their real-life family uses and how many miles they travel using that amount of gas. Show students how to calculate how many miles per gallon a car gets. Ask them to figure out how many miles per gallon their cars get. If a student does not have a car, tell him/her to use a classmate's log to do the calculations. Then allow the class to work together to rank students' cars in order from best gas mileage to worst gas mileage.

9. Make a class graph that shows what kinds of vehicles each student's family owns. These can be divided into groups by vehicle models such as pickup trucks, motorcycles, two-door cars, four-door sedans, station wagons, sport utility vehicles, and sports cars, or they can be classified by the names of vehicle makers.

10. Encourage students to design the cars of their dreams. Allow them to use their imaginations to draw diagrams for any kinds of cars they think they would like to drive. Students may enjoy making models of their dream cars. Invite volunteers to share their diagrams or models.

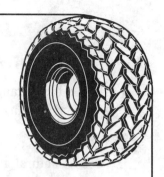

Transportation Price List

NEW CARS

Type of Car	Total Price	Car Payments	Car Insurance
Economy Car	$7,000	$225 per month	$100 per month
Mid-Priced Car	$25,000	$340 per month	$125 per month
Luxury Car	$60,000	$1000 per month	$250 per month

USED CARS

Condition of Car	Total Price	Car Payments	Car Insurance
Fair Condition	$2,500	$65 per month	$50 per month
Good Condition	$3,000	$90 per month	$70 per month
Excellent Condition	$5,000	$100 per month	$90 per month

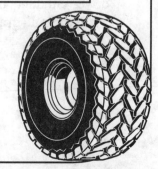

56

Transportation Summary

Use the Transportation Price List (page 56) to fill out summary information for each car your family has.

CAR 1

Type or condition of car: _____

Make: _____

Model: _____

Year: _____

The total price of car: _____

Monthly car payment: _____

Monthly car insurance premium: _____

Weekly cost of gasoline (18 gallons per week): _____

Monthly cost of gasoline (4 x weekly cost): _____

Total spent per month on car 1: _____

CAR 2

Type or condition of car: _____

Make: _____

Model: _____

Year: _____

The total price of car: _____

Monthly car payment: _____

Monthly car insurance premium: _____

Weekly cost of gasoline (18 gallons per week): _____

Monthly cost of gasoline (4 x weekly cost): _____

Total spent per month on car 2: _____

My Car

Use the space below to draw a picture of your car.

58

Other Expenses

Objectives

Students will determine which of two stores has the better prices for individual products as well as the total cost of several products. They will calculate how much money they will spend per month and per year on their personal and household expenses.

Suggested Vocabulary

personal/household supplies, entertainment, appliances, travel, tutor, sales tax, income tax, social security, health insurance, life insurance, homeowner's/renter's insurance

Materials

- Comparison Shopping (page 61), one per student
- Other Expenses Price List (pages 62 and 63), one per student or pair of students
- Other Expenses Summary (pages 64–66), one per student
- Family Profile Record Sheets (page 10) from previous simulations

Optional:

- Basic Operations with Whole Numbers (pages 81 and 82)
- Understanding Decimals (page 83)
- Adding and Subtracting Decimals (page 84)
- Understanding Fractions and Mixed Numbers (page 85)
- Multiplying Fractions and Mixed Numbers (page 86)
- Understanding Percentages (page 87)
- Finding Percentages (page 88)
- Answer Key (page 93)
- Measurement Conversion Chart (page 96)

Procedure

1. Encourage students to brainstorm a list of other expenses that they might have on a regular basis. Write their suggestions on the chalkboard. Review the list, and ask them to determine which items are essential and which ones are not.

2. Work with students to have them calculate the costs of clothing, personal supplies, household extras, and taxes for their families. Prices and tax rates are given on the Other Expenses Price List. Remind students to add 8% sales tax to any subtotal for items that they purchase. Tell them to record their totals and list or describe the kinds of things they buy on their Other Expenses Summary forms. Students may want to look at advertisements or catalogs to write their lists or descriptions.

3. Students can pay for appliances in two ways. They can pay for items in full, using money in their savings or checking accounts, or they can make monthly payments that are 10% of the total cost. If students choose to pay the entire amount for their appliances, make certain that they enter a value of $0.00 for their monthly cost of appliances. Students may choose to pay for some appliances in full and have monthly payments on others.

Other Expenses *(cont.)*

Procedure *(cont.)*

4. Have students determine the amount of money they would have to pay in taxes, based on their monthly incomes.

5. Ask students to figure out a grand total for the amount of money they will spend on other expenses as defined by the Other Expenses Price List.

6. Finally, remind each student to record the monthly and yearly estimates for other expenses on his or her Family Profile Record Sheets, (page 10).

Follow-Up Activities

1. Have students create a shopping list of ten personal items they want to buy. Take them on a field trip to a store. Ask them to price the items on their lists. Discuss whether the prices were more or less than they anticipated.

2. Invite a travel agent to speak to your class. Ask the agent to tell about the kinds of expenses students could expect to pay while on vacation.

3. Divide the class into cooperative learning groups. Provide students with a variety of travel brochures, or have them obtain their own. Have each group plan a vacation. Tell students to work together to determine the location, the length of time spent there, and the amount of money needed to make the trip. Allow time for the groups to share their information.

4. Encourage students to do research to learn how tax money is spent in your community.

5. Show students how to make circle graphs. Tell them to use circle graphs to show what percentage of their incomes are spent on food, housing, transportation, and other expenses.

6. Obtain a copy of a Form 1040 (used to file personal income tax) from the Internal Revenue Service. Reproduce the form and give each student a copy. Discuss the different parts of the form. You may wish to invite an IRS representative or accountant to discuss who must file taxes and how they are filed.

7. Divide the class into two teams. First have students do research to learn about the pros and cons of having either a state or a federal income tax. Then allow them to debate the issue.

8. Have your class conduct a school survey to determine what types of pets are the most popular. Ask students to use the data they obtain to create a table and a graph.

9. Have students conduct interviews to find out about health, life, and homeowners'/renters' insurance policies. Ask them to present their findings to the class.

Comparison Shopping

Other expenses include the personal supplies that you buy and use on a regular basis. Toothpaste, soap, toilet paper, and cologne are a few examples. The prices listed for each product in the following table are the same brand at both stores. First calculate the price per unit for each product. (Remember: price per unit = price ÷ quantity or amount.) The first one is done for you. Then decide which store has the better buy. Finally, add all the prices together to determine which store has the lower total cost.

Brand Name/Product	Price at Alpha Store	Price Per Unit	Price at Beta Store	Price Per Unit	Store with Better Buy
1. Bristles Toothbrush	1 for $0.99	$0.99	*2 for $3.00	$1.50	Alpha Store
2. Silky Style Shampoo	$1.99 for 10 oz.		$2.69 for 12 oz.		
3. Odorama Deodorant	*2 for $3.00		1 for $1.59		
4. Cure-All Antacid	$3.99 for 12 oz.		$6.99 for 24 oz.		
5. Shine It Shoe Polish	*2 for $2.29		*2 for $1.99		
Total Cost for All Products					

*Use the price per unit for these items for Total Cost for All Products.

Now use newspaper advertisements to compare the prices of three personal products that you like to use. Decide which store has the better buy.

Brand Name/Product	Price at _____	Price Per Unit	Price at _____	Price Per Unit	Store with Better Buy
1.					
2.					
3.					
Total Cost for All Products					

Other Expenses Price List

Clothing		
Person	**Time Period**	**Cost**
Adult	Per Month	$100
Child	Per Month	$50

Personal Supplies		
Person	**Time Period**	**Cost**
Adult	Per Week	$50
Child (Age 10 or Older)	Per Week	$30
Child (Under Age 10)	Per Week	$20

Appliances	
Item	**Cost**
Small Color Television	$200
Large Color Television	$1,000
Video Recorder	$250
Washing Machine	$600
Dryer	$300
Dishwasher	$500
Refrigerator	$850
Range (Stove/Oven)	$500
Vacuum Cleaner	$200
Computer	$1,200

Taxes	
Type of Tax	**Percent Taken Out of Each Paycheck**
Federal Income Tax	10%
State Income Tax (if applicable)	2%
Social Security	5%

62

Other Expenses Price List *(cont.)*

EXTRAS

Entertainment		
Type of Entertainment	**Person**	**Cost** (Children under age 3 are free.)
Movie	Per Adult	$7.00
	Per Child	$3.50
Play at a Community Theater	Per Person	$20.00
Sporting Event	Per Person	$10.00
Miniature Golf/Bowling	Per Person	$10.00
Amusement Park	Per Adult	$25.00
	Per Child	$15.00
Circus/Zoo	Per Adult	$15.00
	Per Child	$7.00

Pets		
Type of Pet	**Time Period**	**Cost**
Dog	Per Month	$100
Cat	Per Month	$50
Bird	Per Month	$25
Fish	Per Month	$10

Hired Help		
Type of Hired Help	**Time Period**	**Cost**
Maid/Gardener	Per Week	$40
Daycare	Per Week	$100
Private Nanny	Per Week	$200

Private Education		
Type of Private Education	**Time Period**	**Cost**
Private School	Per Month	$500
Private Tutoring	Per Session	$40

Trips		
Location	**Time Period**	**Cost per Person**
In State	Per Week	$700
Out of State	Per Week	$2,000
Out of the Country	Per Week	$4,000

Other Expenses Summary

Clothing	

Number of adults in my family: _____

Number of Adults x $100 =	Cost per Month

Description of clothing purchased:_____

_____

_____

Number of children in my family: _____

Number of Children x $50 =	Cost per Month

Description of clothing purchased:_____

_____

_____

Subtotal (adults' cost per month + children's cost per month)	
Tax (8% of subtotal [0.08 x subtotal])	
Total (subtotal + tax)	

Taxes		
Type of Tax	**Calculation of Tax**	**Monthly Taxes**
Federal Income Tax	10% of monthly income 0.10 x (monthly income) =	
State Income Tax (if applicable)	2% of monthly income 0.02 x (monthly income) =	
Social Security	5% of monthly income 0.05 x (monthly income) =	

Other Expenses Summary *(cont.)*

Personal Supplies
(deodorant, toothpaste, toys, school supplies, etc.)

Number of adults in my family: _____

(Number of Adults) x $50 x 4 =	Cost per Month

List of items purchased for adults: _____

_____

_____

Number of children (age 10 or older) in my family: _____

(Number of Children Age 10 or Older) x $30 x 4 =	Cost per Month

Number of children (under age 10) in my family: _____

(Number of Children Under Age 10) x $20 x 4 =	Cost per Month

List of items purchased for children: _____

_____

_____

Subtotal (adults' cost per month + children's cost per month)	
Tax (8% of subtotal [0.08 x subtotal])	
Total (subtotal + tax)	

 #192 Math Simulations

Other Expenses Summary *(cont.)*

Appliances	
Cost of Appliances	
Tax (8% of cost [0.08 x cost])	
Total (Cost + Tax)	
Monthly Cost (Total ÷ 12)	

List of appliances purchased: _____

_____

Extras	
Type of Cost	Cost per Month
Entertainment	
Pets	
Hired Help	
Private Education	
(Cost of Trips) ÷ 12	
Total (Add all of the above.)	

List of extras: _____

_____

Total Monthly Cost of Clothing (page 64)	
Total Monthly Cost of Taxes (page 64)	
Total Monthly Cost of Personal Supplies (page 65)	
Total Monthly Cost of Appliances (page 66)	
Total Monthly Cost of Extras (page 66)	
GRAND TOTAL FOR OTHER EXPENSES (Add together all of the above totals.)	

Unexpected Events

Objectives

Students will determine whether unexpected events have positive or negative consequences. They will show how their finances are affected by these events.

Suggested Vocabulary

natural disaster, increase, decrease, laid off, bonus, promotion, prescription medicine, over-the-counter medicine, medical emergency, crime, victim, accident, property damage, injury, inheritance, profit, loss, financial reward

Materials

- Unexpected Event Cards (pages 68–73), enough for at least one card per student on cardstock or heavy paper
- Family Profile Record Sheets (page 10) from previous simulations

Optional:
- Basic Operations with Whole Numbers (pages 81 and 82)
- Understanding Decimals (page 83)
- Adding and Subtracting Decimals (page 84)
- Understanding Fractions and Mixed Numbers (page 85)
- Multiplying Fractions and Mixed Numbers (page 86)
- Understanding Percentages (page 87)
- Finding Percentages (page 88)
- Answer Key (page 93)
- Measurement Conversion Chart (page 96)

Procedure

1. On the chalkboard, make a two-column chart with the headings Positive and Negative. Ask students to brainstorm one list of events that could affect their finances positively and another list of events that could affect their finances negatively. Write students' suggestions on the chart under the appropriate headings.

2. Use cardstock or heavy paper to reproduce the cards shown on pages 68–73. Cut apart the cards, shuffle, and place them facedown in a stack.

3. Have students pick events and read them to the class. Tell students to record the effects of the events on the Family Profile Record Sheet (page 10).

Follow-Up Activities

1. Invite a Red Cross volunteer to speak to your class about disaster relief. Ask how students can volunteer to help the local Red Cross.

2. Have students create a time line that shows major disasters around the world. Ask them to do research to learn about what kind of help is available to victims of disasters.

Unexpected Event Cards

You win a new car on a game show, so you sell your old one for $1,500.

You pay the Internal Revenue Service $1,200 for last year's income tax.

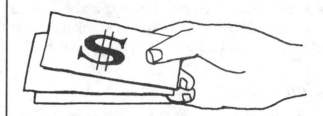

You receive a $100 reward for returning a lost wallet.

You go to the emergency room at the hospital and get stitches. The bill is $175.

You earn a $1,000 bonus from your boss.

Susan Jones 1309 1st Street Huntington Village, CA 92000 (700) 801-9901	331 *May 12* 19 *97* 98-7170/3341

Pay to the order of ___*Joe Green*___ $*1000*___

___*One thousand and* ⁿᵒ/cents___ Dollars

Bank of Huntington Beach
Huntington Beach, California 92647

For ___ *Susan Jones*

⑆341971401⑆ 3940561278⑈

A plumber fixes the leaky faucet in your kitchen for $115.

You receive $200 for your birthday.

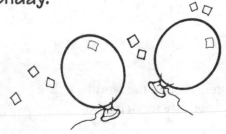

Your mechanic charges you $480 for car repairs.

Unexpected Event Cards *(cont.)*

Your boss gives you a raise of $1,000 per year. 	You spend $100 for prescription medicines. 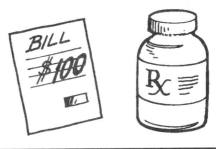
A flood destroys some furniture. It costs $2,000 to purchase new furniture. 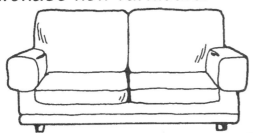	You pay $1,000 for legal expenses. 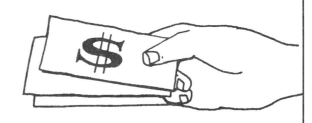
You get a $5,000 pay cut per year. 	You repair your roof for $3,000. 
You need to replace your garage door. It costs $500. 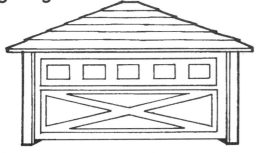	You get a tax refund for $250. 

Unexpected Event Cards *(cont.)*

Your air conditioner/heater breaks down. You pay $500 to repair it.	You cash a savings bond for $100. 
You inherit $5,000 from a great aunt who died. 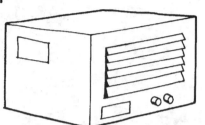	Your pet is sick, and you take it to the vet. It costs $125. 
You buy a computer. It costs $1,200. 	You had a minor car accident. The repairs will cost $300. You pay the $200 deductible, and your insurance company pays the rest. 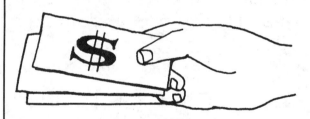
Your dentist tells you that you need major dental work. You pay him $2,500. 	You work a second job and earn $10,000 in one year. 

Unexpected Event Cards *(cont.)*

You spend $2,500 to visit a sick relative. 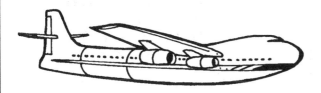	Your stolen wallet is returned, but the $125 you had in it is missing. 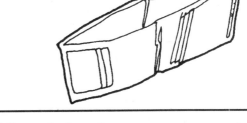
You forget to add a deposit into your checking account. You discover you have $300 more than you thought you did. 	You visit the emergency room for a broken arm. The bill is $1,000. 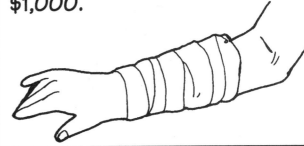
You sell some stocks and make a $200 profit. 	You forgot to pay last month's bills. The penalties on this month's bills add up to $90. 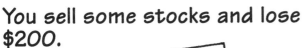
You sell some stocks and lose $200. 	Your refrigerator breaks down. A new one costs you $900. 

Unexpected Event Cards (cont.)

You have a flat tire. It costs you $60 to replace. 	A friend borrows $100 and never pays you back. 
You find $25 in an old wallet that you are about to throw away. 	A magazine pays you $150 for a poem that you wrote. 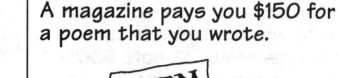
Your car is in the repair shop. You must take a taxi to and from work. Together the trips cost you $50. 	You lose $20 while jogging. It must have fallen out of your pocket. 
You earn $50 selling some old records and CD's. 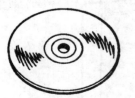	A friend is collecting funds for your community's homeless. You donate $50. 

Unexpected Event Cards *(cont.)*

You are parked illegally and get a ticket. The fine is $60.

You need glasses for the first time in your life. They cost you $125.

You repair your broken television for $80.

Your favorite football team is finally playing in the Super Bowl. You pay $100 for a ticket.

You break a window in your living room. It costs you $70 to replace.

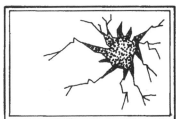

You get a $25 rebate on a new bed that you bought.

You always stick extra change in your piggy bank. Today you count it and find that you have saved $50.

You are a witness to a crime and work with the police to help them catch the criminal. You earn a $100 reward.

Total Expenses

Objectives

Students will calculate their total expenses per month and per year. Then they will decide whether their incomes are sufficient for the amount of expenses they have. Students will brainstorm a list of ways to save money if they find they are living beyond their means.

Suggested Vocabulary

net income, gross income, expenses, budget, social security, federal/state income tax, deductions, withholding

Materials

- Can I Pay My Bills? (pages 77–80), one copy per student
- Family Profile Record Sheets (page 10) from previous simulations

Optional:

- Basic Operations with Whole Numbers (pages 81 and 82)
- Understanding Decimals (page 83)
- Adding and Subtracting Decimals (page 84)
- Understanding Fractions and Mixed Numbers (page 85)
- Multiplying Fractions and Mixed Numbers (page 86)
- Understanding Percentages (page 87)
- Finding Percentages (page 88)
- Answer Key (page 93)
- Measurement Conversion Chart (page 96)

Procedure

1. Students need to determine if they have enough money to cover all of their expenses. The first step is to see how much money their family is earning after taxes have been deducted. Discuss with students the purpose of taxes and why the salary people earn is more than the money they actually take home in their paychecks. Explain the difference between gross and net incomes.

2. Have students calculate 10% for federal income tax, 2% for state tax (if applicable), and 5% for social security. Remind them to figure out their taxes based on their total monthly salaries, not their total yearly salaries. Be sure students use the same monthly totals for each tax calculation.

3. Tell students to determine the total amounts they must deduct from their incomes for taxes. Have them subtract the total amounts of taxes from their monthly incomes.

4. You may wish to provide copies of a tax table from the Internal Revenue Service. Allow students to practice finding the amount of tax people pay based on the simulated family sizes and incomes represented by students in your class.

Total Expenses *(cont.)*

Procedure (cont.)

5. Have students review how much they estimated spending for food, housing, transportation, other expenses, and unexpected events, using their Family Profile Record Sheets (page 10). Then ask them to calculate their total monthly expenses on their profiles.

6. Some students may wish to pay for some of their expenses with money they have in savings. They can choose to use their savings to make a down payment on a house or a car, to purchase appliances, take trips, etc. Remind students that they must have sufficient funds in their savings accounts if they plan to purchase things using these monies. Otherwise, they will need to use their monthly incomes for these expenses as well as all other expenses that they incur.

7. Invite students to share experiences they have had with their own savings accounts. Ask them how they earn money to put in savings.

8. To determine if they can pay their bills, students need to subtract the total monthly expenses from the monthly incomes after taxes are deducted. Provide students with copies of Can I Pay My Bills? so they can make these calculations.

9. Explain that if the difference is a negative number, this means they are spending more money than they are earning each month. If the difference is a positive number, this means they are earning more than they are spending so they will have money left over at the end of each month. Point out that this extra money can be placed in savings.

10. Discuss what it means to have an account that is overdrawn. Point out that banks have penalties (fines) for people who are overdrawn.

11. Encourage students to share their results.

12. Ask students to compare their predicted expenses with their estimated expenses.

13. Divide the class into cooperative learning groups. Ask them to brainstorm a list of ways to cut family expenses so that the bills can be paid. Have students record their ideas.

14. Have students write essays to tell what they have learned about family finances by using the simulations.

Total Expenses *(cont.)*

Follow-Up Activities

1. Have students write about one or more of the following:

   - a description of each simulation with an explanation of their individual results
   - a comparison of their predicted and estimated expenses based on the Family Profile Record Sheet (page 10)
   - an explanation of how an unexpected event affected their families
   - personal reactions and thoughts about the different simulations presented in this book
   - suggestions for other things they would like to have included in the simulations

2. Have students make booklets to show their work from the simulations. Encourage them to share their booklets with their families.

3. Ask students to brainstorm a list of other things on which they might need to spend money. Write their suggestions on the chalkboard. Based on their experiences, have them estimate how much money each item might cost. If you prefer, they can do research at local stores or use advertisements to determine the actual cost of each item.

4. Help students prepare presentations for parents and/or other visitors. Ask them to tell about the process they used to determine whether they had enough money to pay their bills. Remind them to tell about their predictions, their estimated expenses, and explain how these numbers compared.

5. Have students whose expenses did not exceed their incomes role-play being financial planners. Ask them to help students who are living beyond their means to make adjustments in their budgets.

6. Invite students to share experiences they have with helping their real-life families manage their finances. Encourage them to find ways to help their parents reduce spending.

7. Discuss different types of investments. Invite a stockbroker to tell students about the advantages and disadvantages of specific investments.

8. Have each student pick a specific stock that is reported in the newspaper. Ask them to track their stocks each day for two weeks. Tell them to record the data in tables and then use the information to create line graphs.

9. Create a bulletin board to display current financial newspaper and magazine clippings that students collect. Allow time for students to read and discuss the articles.

Can I Pay My Bills?

Complete the calculations below to determine how much money you have available to pay your monthly bills.

Gross Income (Before Taxes Are Deducted)	
	Amount of Income
My Yearly Income	
My Spouse's Yearly Income (if applicable)	
My Family's Total Yearly Income (my yearly income + my spouse's yearly income)	
My Family's Total Monthly Gross Income (total yearly income ÷ 12)	

Taxes	
	Amount of Tax per Month
Federal Income Tax (10% of total monthly income [0.10 x total monthly income])	
State Income Tax (if applicable) (2% of total monthly income [0.02 x total monthly income])	
Social Security (5% of total monthly income [0.05 x total monthly income])	
Total Taxes (federal + state + social security)	

Net Income (After Taxes Are Deducted)	
	Amount of Income
My Family's Total Monthly Net Income (total monthly gross income – total taxes)	

Can I Pay My Bills? *(cont.)*

Savings	
Starting amount in my family's savings account: _____	
	Amount Spent
Housing	
Transportation	
My Car	
My Spouse's Car (if applicable)	
Appliances	
Trips	
Unexpected Events	
Other	
Total Amount of Savings Used (Add all of the above.)	
Final Amount Left in Savings (starting amount in savings – total money spent)	

Describe some ways that people can increase their savings.

_____

_____

_____

_____

_____

_____

_____

_____

_____

_____

78

Can I Pay My Bills? *(cont.)*

Expenses		Amount
Monthly Food Expenses		
Monthly Housing Expenses		
Monthly Transportation Expenses		
Other Monthly Expenses		
Expenses for Unexpected Events		
Total Monthly Expenses (Add all of the above.)		

My family's monthly income after taxes are withdrawn:_____

My family's total monthly expenses: _____

> Now subtract your family's total monthly expenses from your family's monthly income after taxes. Show your work here.

Can you pay your monthly bills? If you cannot pay your monthly bills, tell some ways that you could reduce your spending. If you can pay your monthly bills, explain how you were able to avoid overspending.

_____

_____

_____

_____

_____

_____

_____

Can I Pay My Bills? *(cont.)*

Write an short essay to tell what you learned about family finances by using the simulations.

80

Basic Operations with Whole Numbers

Helpful Hints: If there are no parentheses, you should solve the problem in this order— multiplication, division, addition, and subtraction. You can remember the order of operations using the first letter of each word to say the greeting, "**M**y **D**ear **A**unt **S**ally."

If you are not sure how to read a large number, use a place value chart like the one shown below.

millions	hundred thousands	ten thousands	thousands	hundreds	tens	ones
5 ,	7	3	2 ,	6	4	1

Write these problems in vertical form, and then solve them. Be sure to show your work.

1. 30 + 500 + 63 + 751 =	**2.** 75 x 80 =	**3.** 1,239 − 784 =	**4.** 646 ÷ 34 =
5. 9,765 x 54 =	**6.** 80,345 − 2,245 =	**7.** 3045 ÷ 15 =	**8.** 67,546 + 8,556 =

Basic Operations with Whole Numbers *(cont.)*

Write these problems in vertical form, and then solve them. Be sure to show your work.

9. 49,320 ÷ 6 =	10. 8,210 x 24 =	11. 9,830 + 497 + 22,980 =	12. 90,000 ÷ 100 =
13. 5,058 x 400 =	14. 40,590 – 2,825 =	15. (1,258 – 827) + 993 =	16. 4,400 ÷ 220 =
17. 6,002 x 847 =	18. 65,538 ÷ 331 =	19. (40,491 – 13) + 600 =	20. 500 x (200 ÷ 25) =

82

Understanding Decimals

Example ➡	6 + 97.81 + 89.4 =

Step 1 Change whole numbers to decimal form. Use place-holding zeros after the decimal.

6 = **6.0**

Step 2 Line up the decimal points when you write the problem in vertical form.

```
  6.0
 97.81
+89.4
------
```

Step 3 Write place-holding zeros so each number in the problem has the same number of decimal places. This will help you keep your numbers correctly aligned.

```
  6.00
 97.81
+89.40
------
```

Step 4 Write the decimal point in the answer directly below the decimal points in the problem.

```
  6.00
 97.81
+89.40
------
    .
```

Step 5 Calculate from right to left. Regroup if necessary.

```
  6.00        1          21          21
 97.81       6.00        6.00        6.00
+89.40      97.81       97.81       97.81
------     +89.40      +89.40      +89.40
    .1     ------      ------      ------
             .21         3.21       193.21
```

Follow the same steps as shown above for a subtraction problem.

Adding and Subtracting Decimals

Write these problems in vertical form, and then add or subtract. Remember to line up the decimal points. Refer to page 83 if you need to review the steps or study an example.

1 6.35 – 2.58 =	**2** 85.3 – 66.47 =	**3** $19 + $56.28 =	**4** 56.16 + 7.90 =
5 0.093 + 0.41 =	**6** 1.002 + 10.2 =	**7** $30 + $0.30 + $3.02 =	**8** $463.21 + $19.10 =
9 $1.04 – $0.22 =	**10** $8.25 + $3.62 =	**11** $45.10 + $62 =	**12** $88 + $1.05 =
13 $40 – $2.03 =	**14** 9.001 – 5.57 =	**15** 85.3 + 47.66 =	**16** 3.65 + 2.89 =

84

Understanding Fractions and Mixed Numbers

Helpful Hint: A numerator is the top number of a fraction. A denominator is the bottom number of a fraction. Notice the word denominator starts with the letter **d**. Think of the word *down* when you see the letter **d**. This will help you remember that the denominator goes "down" to the bottom of the fraction while the numerator goes "up" to the top.

Example 1 ➝	$\frac{1}{2} \times \frac{2}{3} =$

Step 1	Multiply the numerators.	$\frac{1}{2} \times \frac{2}{3} = \frac{2}{}$
Step 2	Multiply the denominators.	$\frac{1}{2} \times \frac{2}{3} = \frac{2}{6}$
Step 3	Simplify the answer to its lowest terms. This is done by dividing the numerator and the denominator by the greatest common factor — GCF.	$\frac{2 \div 2}{6 \div 2} = \frac{1}{3}$ (The GCF is 2.)

Example 2 ➝	$\frac{3}{4} \times 8 =$

Step 1	Write the whole number as a fraction.	$8 = \frac{8}{1}$
Step 2	Multiply the numerators.	$\frac{3}{4} \times \frac{8}{1} = \frac{24}{}$
Step 3	Multiply the denominators.	$\frac{3}{4} \times \frac{8}{1} = \frac{24}{4}$
Step 4	Simplify the answer to its lowest terms, using the GCF.	$\frac{24 \div 4}{4 \div 4} = \frac{6}{1}$ (The GCF is 4.)
Step 5	If your answer is an improper fraction (the numerator is greater than the denominator), rewrite it as a whole number or a mixed number.	$\frac{6}{1} = 6$

Multiplying Fractions and Mixed Numbers

Multiply. Then simplify each answer to its lowest terms. Refer to page 85 if you need to review the steps or study the examples.

1 $\frac{1}{4} \times 3\frac{1}{3} =$ _____	**2** $\frac{1}{2} \times \frac{2}{5} =$ _____	**3** $7\frac{3}{4} \times 1\frac{1}{2} =$ _____
4 $\frac{1}{2} \times \frac{4}{5} =$ _____	**5** $2\frac{1}{4} \times \frac{1}{5} =$ _____	**6** $\frac{1}{2} \times 8 =$ _____
7 $\frac{3}{5} \times 15 =$ _____	**8** $\frac{2}{3} \times 6 =$ _____	**9** $\frac{3}{8} \times 3\frac{1}{4} =$ _____
10 $\frac{4}{5} \times 40 =$ _____	**11** $3\frac{1}{10} \times \frac{5}{6} =$ _____	**12** $\frac{2}{3} \times \frac{2}{5} =$ _____
13 $\frac{1}{3} \times 1\frac{6}{7} =$ _____	**14** $\frac{3}{4} \times \frac{1}{6} =$ _____	**15** $5 \times 6\frac{2}{3} =$ _____

Understanding Percentages

Helpful Hint: In mathematics, the word *of* tells you to multiply.

Example 1 ➞	22% of 38 = _____

Step 1	Rewrite the equation as a multiplication problem.	22% x 38 =
Step 2	Rename the percent as a decimal by removing the percent symbol and moving the decimal point two places to the left.	0.22 x 38 =
Step 3	Multiply to solve the problem.	0.22 x 38 = 8.36

Example 2 ➞	5% of 6.9 = _____

Step 1	Rewrite the equation as a multiplication problem.	5% x 6.9 =
Step 2	Rename the percent as a decimal.	0.05 x 6.9 =
Step 3	Multiply to solve the problem	0.05 x 6.9 = 0.345

Example 3 ➞	130% of 8 = _____

Step 1	Rewrite the equation as a multiplication problem.	130% x 8 =
Step 2	Rename the percent as a decimal.	1.30 x 8 =
Step 3	Multiply to solve the problem.	1.30 x 8 = 10.4

Finding Percentages

Solve for the percentage. Refer to page 87 if you need to review the steps or study some examples.

1 4% of 37 =	2 50% of 6 =	3 25% of 57 =	4 55% of 64 =
5 90% of 20 =	6 70% of 300 =	7 34% of 5.7 =	8 7.4% of 200 =
9 5.6% of 283 =	10 48% of 860 =	11 85% of 3000 =	12 40% of 120 =
13 75% of 10 =	14 8% of 15 =	15 5% of 6.9 =	16 12% of 64 =

Understanding Perimeter

To find the perimeter of any polygon, add together the lengths of the sides. Examples of when you might want to know the perimeter include building a fence around a garden, putting a frame around a picture, and determining the distance around a baseball diamond.

Helpful Hint: Notice the word *rim* in *perimeter*. To find the perimeter, add the lengths of the sides as you "walk around the rim."

Examples

1.

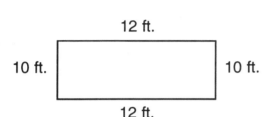

12 ft.
10 ft.   10 ft.
12 ft.

```
  12
  10
  12
+ 10
  44
```

Perimeter =
44 feet

2.

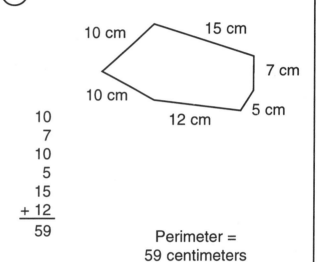

10 cm   15 cm
                    7 cm
10 cm
         12 cm   5 cm

```
  10
   7
  10
   5
  15
+ 12
  59
```

Perimeter =
59 centimeters

3.

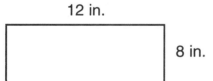

12 in.

8 in.

Opposite sides of a rectangle are the same length. Write the missing labels before adding together the lengths of the sides.

```
  12
   8
  12
+  8
  40
```

Perimeter =
40 inches

4.

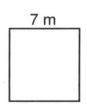

7 m

All sides of a square are the same length. Write the missing labels before adding together the lengths of the sides.

```
   7
   7
   7
+  7
  28
```

Perimeter =
28 meters

Finding Perimeters

Find the perimeters. Refer to page 89 if you need some examples.

(1.) 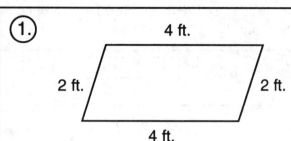 Perimeter = _____	**(2.)** 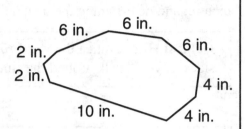 Perimeter = _____
(3.) 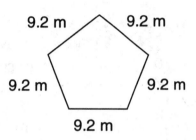 Perimeter = _____	**(4.)** 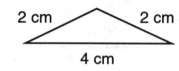 Perimeter = _____
(5.) You need to build a fence around your rectangular vegetable garden. The length is 16 meters and the width is 18 meters. What is the perimeter? 18 m 16 m _____	**(6.)** Mr. Thomas is making a picture frame for his daughter's picture. The dimensions of the picture are 8 inches by 6 inches. What is the perimeter of the picture? 6 in. 8 in. _____
(7.) Ms. Loa wants to put a decorative border on her bulletin board. How much border does she need if the bulletin board is 6 feet by 3 feet? 6 ft. 3 ft. _____	**(8.)** Rami and Alex take their dog for a walk around the jogging track three times. The length of the track is 100 meters and the width is 50 meters. How far did they walk? 100 m 50 m _____

Understanding Area

To find the area of any square or rectangle, multiply the length times the width (A = l x w). To find the area of a triangle, multiply ½ times the base times the height (A = ½ [b x h]). Examples of when you might want to know the area include installing carpet, covering a wall with wallpaper, and finding out how large a football field is. Area is written as square units or units² (units squared).

Example 1

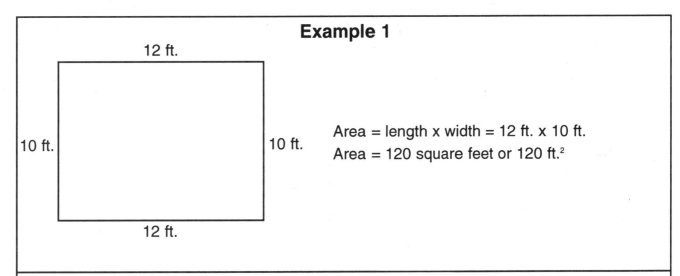

Area = length x width = 12 ft. x 10 ft.

Area = 120 square feet or 120 ft.²

Example 2

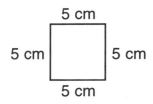

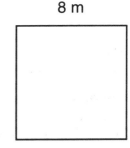

Area = length x width = 5 cm x 5 cm

Area = 25 square centimeters or 25 cm²

Area = length x width = 8 m x 8 m

Area = 64 square meters or 64 m²

Example 3

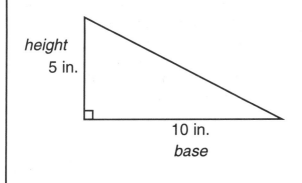

Area = ½ (base x height) = (base x height) ÷ 2

Area = (10 in. x 5 in.) ÷ 2 = 50 in. ÷ 2

50 in. ÷ 2 = 25 square inches or 25 in.²

Finding Areas

Find the area of each figure. Refer to page 91 if you need some examples.

1.

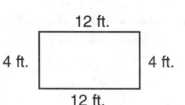

12 ft.

4 ft.   4 ft.

12 ft.

Area = _____

2.

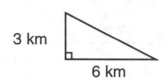

3 km

6 km

Area = _____

3.

13 yd.

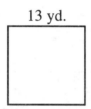

Area = _____

4.

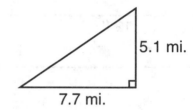

5.1 mi.

7.7 mi.

Area = _____

5. Danny wants to line the bottom of his bird's cage with clean paper. He knows the length is 30 cm and the width is 18 cm. What is the area?

30 cm

18 cm

_____

6. Candace is helping her mother hang wallpaper on one wall which measures 12 meters by 15 meters. What is the area of the wall?

12 m

15 m

_____

7. Mr. Shapiro wants to put new liners on his four cabinet shelves. Each shelf is 24 inches by 14 inches. What is the area of each shelf? What is the combined area of all four shelves?

24 in.

14 in.

_____

8. Nikita has made a glass coaster that is in the shape of a triangle. She wants to cover the top of it with blue contact paper. If the base of the triangle is 11 centimeters and the height is 8 centimeters, how much contact paper does she need?

8 cm

11 cm

_____

Answer Key

Page 61

Brand Name/Product	Price at Alpha Store	Price Per Unit	Price at Beta Store	Price Per Unit	Store with Better Buy
1. Bristles Toothbrush	1 for $0.99	$0.99	*2 for $3.00	$1.50	Alpha Store
2. Silky Style Shampoo	$1.99 for 10 oz.	$0.19 per oz.	$2.69 for 12 oz.	$0.22 per oz.	Alpha Store
3. Odorama Deodorant	*2 for $3.00	$1.50	1 for $1.59	$1.59	Alpha Store
4. Cure-All Antacid	$3.99 for 12 oz.	$0.33 per oz.	$6.99 for 24 oz.	$0.29 per oz.	Beta Store
5. Shine It Shoe Polish	*2 for $2.29	$1.14	*2 for $1.99	$1.00	Beta Store
Total Cost for All Products	$9.61		$13.77		Alpha Store

Page 81
1. 1,344
2. 6,000
3. 455
4. 19
5. 527,310
6. 78,100
7. 203
8. 76,102

Page 82
9. 8,220
10. 197,040
11. 33,307
12. 900
13. 2,023,200
14. 37,765
15. 1,424
16. 20
17. 5,083,694
18. 198
19. 41,078
20. 4,000

Page 84
1. 3.77
2. 18.83
3. $75.28
4. 64.06
5. 0.503
6. 11.202
7. $33.32
8. $482.31
9. $0.82
10. $11.87
11. $107.10
12. $89.05
13. $37.97
14. 3.431
15. 132.96
16. 6.54

Page 86
1. $^{10}/_{12} = {}^5/_6$
2. $^2/_{10} = {}^1/_5$
3. $^{93}/_8 = 11\,{}^5/_8$
4. $^4/_{10} = {}^2/_5$
5. $^9/_{20}$
6. $^8/_2 = 4$
7. $^{45}/_5 = 9$
8. $^{12}/_3 = 4$
9. $^{39}/_{32} = 1\,{}^7/_{32}$
10. $^{160}/_5 = 32$
11. $^{155}/_{60} = 2\,{}^7/_{12}$
12. $^4/_{15}$
13. $^{13}/_{21}$
14. $^3/_{24} = {}^1/_8$
15. $^{100}/_3 = 33\,{}^1/_3$

Page 88
1. 1.48
2. 3
3. 14.25
4. 35.2
5. 18
6. 210
7. 1.938
8. 14.8
9. 15.848
10. 412.8
11. 2,550
12. 48
13. 7.5
14. 1.2
15. 0.345
16. 7.68

Page 90
1. 12 ft
2. 40 in.
3. 46 m
4. 8 cm
5. 68 m
6. 28 in.
7. 18 ft.
8. 900 m

Page 92
1. 48 square feet (48 ft.^{2})
2. 9 square kilometers (9 km^2)
3. 169 square yards (169 yd.^{2})
4. 19.635 square miles (19.64 mi.^{2})
5. 540 square centimeters (540 cm^2)
6. 180 square meters (180 m^2)
7. 336 square inches (336 in.^{2}); 1,344 square inches (1,344 in.^{2})
8. 44 square centimeters (44 cm^2)

Centimeter Rulers

Make enough transparencies of this page so each student will get a ruler. Cut out the rulers and distribute them. Allow students to use the transparent rulers for measuring in centimeters. You may also wish to have them practice using inch rulers (page 95).

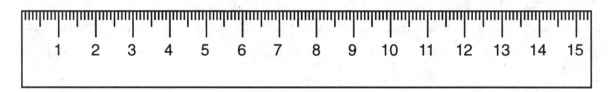

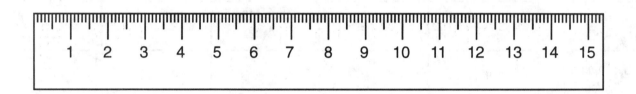

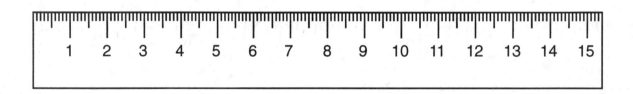

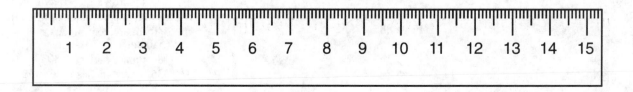

Inch Rulers

Make enough transparencies of this page so each student will get a ruler. Cut out the rulers and distribute them. Allow students to use the transparent rulers for measuring in inches. You may also wish to have them practice using centimeter rulers (page 94).

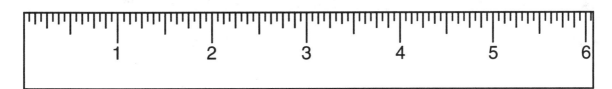

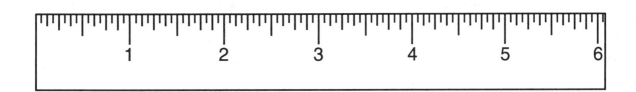

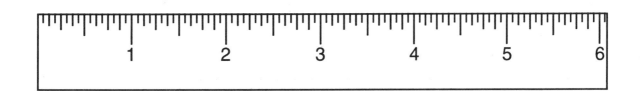

Measurement Conversion Chart

	Metric	Customary
Length	1 centimeter (cm) = 10 millimeters (mm) 1 meter (m) = 100 centimeters (cm) 1 kilometer (km) = 1,000 meters (m)	1 foot (ft.) = 12 inches (in.) 1 yard (yd.) = 3 feet (ft.) 1 mile (mi.) = 5,280 feet (ft.) 1 mile (mi.) = 1,760 yards (yd.)
Volume	1 liter (L)= 1,000 milliliters (mL)	1 cup (c.) = 8 ounces (oz.) 1 pint (pt.) = 2 cups (c.) 1 quart (qt.) = 2 pints (pt.) 1 gallon (gal.) = 128 ounces (oz.) 1 gallon (gal.) = 4 quarts (qt.)
Weight	1 gram (g) = 1,000 milligrams (mg) 1 kilogram (kg) = 1,000 grams (g)	1 pound (lb.) = 16 ounces (oz.) 1 ton = 2,000 pounds (lb.)

	*Changing Customary to Metric	*Changing Metric to Customary Length
Length	1 inch = 2.54 centimeters 1 foot = 30 centimeters 1 yard = 0.91 meters 1 mile = 1.6 kilometers	1 centimeter = 0.4 inches 1 meter = 1.09 yards 1 kilometer = 0.62 miles
Volume	1 cup = 240 milliliters 1 pint = 0.47 liters 1 quart = 0.95 liters 1 gallon = 3.79 liters	1 liter = 1.06 quarts 1 liter = 0.26 gallons
Weight	1 ounce = 28.4 grams 1 pound = 0.45 kilograms	1 gram = 0.035 ounces 1 kilogram = 2.21 pounds

* Approximations